Basic Computation

Working with Decimals

Loretta M. Taylor, Ed. D.
Mathematics Teacher
Hillsdale High School
San Mateo, California

Harold D. Taylor, Ed. D.
Head, Mathematics Department
Aragon High School
San Mateo, California

DALE
SEYMOUR
PUBLICATIONS
P.O. BOX 10888
PALO ALTO, CA 94303

ABOUT THE PROGRAM

WHAT IS THE BASIC COMPUTATION LIBRARY?

The books in the BASIC COMPUTATION library together provide comprehensive practice in all the essential computational skills. There are practice books and a test book. The practice books consist of carefully sequenced drill worksheets organized in groups of five. The test book contains daily quizzes (160 quizzes in all), semester tests, and year-end tests written in standardized-test formats.

If you find this book effective, you may want to use others in the series. Build your own library to suit your own needs.

BOOK 1	WORKING WITH WHOLE NUMBERS
BOOK 2	UNDERSTANDING FRACTIONS
BOOK 3	WORKING WITH FRACTIONS
BOOK 4	WORKING WITH DECIMALS
BOOK 5	WORKING WITH PERCENTS
BOOK 6	UNDERSTANDING MEASUREMENT
BOOK 7	FINDING AREA AND PERIMETER
BOOK 8	WORKING WITH CIRCLES AND VOLUME
BOOK 9	APPLYING COMPUTATIONAL SKILLS
TEST BOOK	BASIC COMPUTATION QUIZZES AND TESTS

WHO CAN USE THE BASIC COMPUTATION WORKSHEETS LIBRARY?

Classroom teachers, substitute teachers, tutors, parents, and persons wishing to study on their own can use these materials. Although written specifically for the general math classroom, books in the BASIC COMPUTATION library can be used with any program requiring carefully sequenced computational practice. The material is appropriate for use with any person, young or old, who has not yet certified computational proficiency. It is especially suitable for middle school, junior high school, and high school students who need to master the essential computational skills necessary for mathematical literacy.

WHAT IS IN THIS BOOK?

This book is a practice book. In addition to these teacher notes, it contains student worksheets, example problems, and a record form.

Worksheets

The worksheets are designed to give even the slowest student a chance to master the essential computational skills. Most worksheets come in five equivalent forms allowing for pretesting, practice, and posttesting on any one skill. Each set of worksheets provides practice in only one or two specific skills and the work progresses in very small steps from one set to the next. Instructions are clear and simple, with handwritten samples of the exercises completed. Ample practice is provided on each page, giving students the opportunity to strengthen their skills. Answers to each problem are included in the back of the book.

Example Problems

Fully-worked examples show how to work each type of exercise. Examples are keyed to the worksheet pages. The example solutions are written in a straightforward manner and are easily understood.

Record Form

A record form is provided to help in recording progress and assessing instructional needs.

Answers

Answers to each problem are included in the back of the book.

HOW CAN THE BASIC COMPUTATION LIBRARY BE USED?

The materials in the BASIC COMPUTATION library can serve as the major skeleton of a skills program or as supplements to any other computational skills program. The large number of worksheets gives a wide variety from which to choose and allows flexibility in structuring a program to meet individual needs. The following suggestions are offered to show how the BASIC COMPUTATION library may be adapted to a particular situation.

Minimal Competency Practice

In various fields and schools, standardized tests are used for entrance, passage from one level to another, and certification of competency or proficiency prior to graduation. The materials in the BASIC COMPUTATION library are particularly well-suited to preparing for any of the various mathematics competency tests, including the mathematics portion of the General Educational Development test (GED) used to certify high school equivalency.

Together, the books in the BASIC COMPUTATION library give practice in all the essential computational skills measured on competency tests. The semester tests and year-end tests from the test book are written in standardized-test formats. These tests can be used as sample minimal competency tests. The worksheets can be used to brush up on skills measured by the competency tests.

Skill Maintenance

Since most worksheets come in five equivalent forms, the computation work can be organized into weekly units as suggested by the following schedule. Day one is for pretesting and introducing a skill. The next three days are for drill and practice followed by a unit test on the fifth day.

AUTHORS' SUGGESTED TEACHING SCHEDULE

	Day 1	Day 2	Day 3	Day 4	Day 5
Week 1	pages 1 and 2 pages 11 and 12	pages 3 and 4 pages 13 and 14	pages 5 and 6 pages 15 and 16	pages 7 and 8 pages 17 and 18	pages 9 and 10 pages 19 and 20
Week 2	pages 21 and 22 pages 31 and 32	pages 23 and 24 pages 33 and 34	pages 25 and 26 pages 35 and 36	pages 27 and 28 pages 37 and 38	pages 29 and 30 pages 39 and 40
Week 3	pages 41 and 42 pages 51 and 52	pages 43 and 44 pages 53 and 54	pages 45 and 46 pages 55 and 56	pages 47 and 48 pages 57 and 58	pages 49 and 50 pages 59 and 60
Week 4	pages 61 and 62 pages 71 and 72	pages 63 and 64 pages 73 and 74	pages 65 and 66 pages 75 and 76	pages 67 and 68 pages 77 and 78	pages 69 and 70 pages 79 and 80

The daily quizzes from BASIC COMPUTATION QUIZZES AND TESTS can be used on the drill and practice days for maintenance of previously-learned skills or diagnosis of skill deficiencies.

A five-day schedule can begin on any day of the week. The authors' ideal schedule begins on Thursday, with reteaching on Friday. Monday and Tuesday are for touch-up teaching and individualized instruction. Wednesday is test day.

Supplementary Drill

There are more than 18,000 problems in the BASIC COMPUTATION library. When students need more practice with a given skill, use the appropriate worksheets from the library. They are suitable for classwork or homework practice following the teaching of a specific skill. With five equivalent pages for most worksheets, adequate practice is provided for each essential skill.

HOW ARE MATERIALS PREPARED?

The books are designed so the pages can be easily removed and reproduced by Thermofax, Xerox, or a similar process. For example, a ditto master can be made on a Thermofax for use on a spirit duplicator. Permanent transparencies can be made by processing special transparencies through a Thermofax or Xerox.

Any system will run more smoothly if work is stored in folders. Record forms can be attached to the folders so that either students or teachers can keep records of individual progress. Materials stored in this way are readily available for conferences.

EXAMPLE PROBLEMS

GRAPHING FRACTIONS

EXAMPLE   Graph $\frac{1}{4}$ and $\frac{5}{9}$.

Solution: First write each fraction as a decimal.

$\frac{1}{4} = 1 \div 4 = 0.25$ and $\frac{5}{9} = 5 \div 9 = 0.\overline{5}$

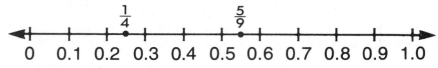

MORE GRAPHING FRACTIONS

EXAMPLE   Graph $\frac{3}{5}$ and $\frac{3}{8}$.

Solution: First write each fraction as a decimal.

$\frac{3}{5} = 3 \div 5 = 0.6$ and $\frac{3}{8} = 3 \div 8 = 0.375$

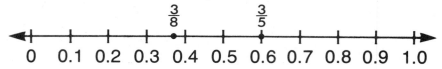

ADDITION OF DECIMALS

EXAMPLE 1 Find the sum:
```
     602.84
      37.3
     157.662
  +   54.89
```

Solution: 852.692

EXAMPLE 2 Add 67.453, 894.42, 51.45, and 87.684.

Solution: Write the numbers in a column.
Keep the decimal points in line
vertically. Then, add.

```
    67.453
   894.42
    51.45
 +  87.684
 ─────────
  1101.007
```

SUBTRACTION OF DECIMALS

EXAMPLE   Find the difference:
```
    803.25
  −  32.73
  ────────
```

Solution: 770.52

EXAMPLE 2 Subtract 12.348 from 25.

Solution: Write the numbers in a column. Keep the decimal points in line vertically. Fill the zeros where needed. Then, subtract.

$$\begin{array}{r} 25.000 \\ -\,12.348 \\ \hline 12.652 \end{array}$$

PLACE VALUE FOR DECIMALS

EXAMPLE   Write 377.438292 in the blanks below.

millions	hundred thousands	ten thousands	thousands	hundreds	tens	ones	and	tenths	hundredths	thousandths	ten-thousandths	hundred-thousandths	millionths
—	—	—	—	—	—	—	—	—	—	—	—	—	—

Solution:

millions	hundred thousands	ten thousands	thousands	hundreds	tens	ones	and	tenths	hundredths	thousandths	ten-thousandths	hundred-thousandths	millionths
—	—	—	—	3	7	7	.	4	3	8	2	9	2

WRITING DECIMALS

EXAMPLE 1 Write the numeral. Fill in zeros where needed.

4 in the thousandths place
6 in the hundreds place
3 in the thousands place
7 in the hundredths place

Solution: The hundreds and thousands places are to the left of the decimal point. The hundredths and thousandths places are to the right of the decimal point.

3 6 0 0 . 0 7 4

EXAMPLE 2 Write the numeral for three million, three hundred seventeen thousand and fifteen hundredths.

Solution: The "and" stands for the decimal point. Fill in zeros where needed.

3,317,000.15

ROUNDING TO THE ONES, TENS, AND HUNDREDS PLACES

EXAMPLE   Round 8465.271 to the ones, tens, and hundreds places.

Solution: Look at the digit in the place to the right of the place being rounded to. If it is 5 or greater than 5, add 1 to the digit in the place being rounded to. Write zeros in the places to the right of the place being rounded to. Drop the decimal point and all digits to the right of it.

8465.271 rounded to the ones place is 8465.
8465.271 rounded to the tens place is 8470.
8465.271 rounded to the hundreds place is 8500.

ROUNDING TO THE TENTHS AND HUNDREDTHS PLACES

EXAMPLE   Round 8465.271 to the tenths and hundredths places.

Solution: Look at the digit in the place to the right of the place being rounded to. If it is 5 or greater than 5, add 1 to the digit in the place being rounded to. Drop all digits to the right of the place being rounded to.

8465.271 rounded to the tenths place is 8465.3.
8465.271 rounded to the hundredths place is 8465.27.

MULTIPLICATION AND DIVISION OF DECIMALS

EXAMPLE 1 Find 73.312×10^2.

Solution: In 10^2, the exponent is 2. Move the decimal point to the right the same number of places as the exponent for 10.

$73.812 \times 10^2 = 7381.2$

EXAMPLE 2 Find 21.324×6.44.

Solution:

21.324	3 decimal places
$\times$ 6.44	+2 decimal places
85296	
85296	
127944	
137.32656	5 decimal places

EXAMPLE 3 Find $76.843 \div 1000$.

    Solution: Move the decimal point to the left the same number of places as there are zeros in 1000.

        $76.843 \div 1000 = 0.076843$

EXAMPLE 4 Find $7.32 \overline{)23.89248}$.

    Solution: Move the decimal point to the right to make the divisor a whole number . Move the decimal point in the dividend to the right the same number of places. Then, divide.

$$
\begin{array}{r}
3.264 \\
7.32\,\overline{)\,23.89248} \\
2196 \\
\overline{1932} \\
1464 \\
\overline{4684} \\
4392 \\
\overline{2928} \\
2928 \\
\overline{0}
\end{array}
$$

COMPARING FRACTIONS AND DECIMALS

EXAMPLE   Use $<, >$, or $=$ to compare $\frac{7}{9}$ and 0.7.

    Solution: The symbol $<$ means *less than.* The symbol $>$ means *greater than.* Write the fraction as a decimal.

    $\frac{7}{9} = 0.777...$ and $0.7 = 0.700 ...$

    Compare digits in each place. In the tenths place, $7 = 7$. In the hundredths place, $7 > 0$. So $\frac{7}{9} > 0.7$.

ORDERING DECIMALS

EXAMPLE   Write 0.17, 0.02, and 0.33 in order from least to greatest.

    Solution: Compare the digits in each place from left to right. Use the first place in which the digits are different to order the numbers.

        ↓   ↓   ↓

    0.02   0.17   0.33

MORE ORDERING DECIMALS

EXAMPLE   Write 2.13, 2.18, and 2.107 in order from the least to greatest.

Solution: Compare the digits in each place from left to right. Use the first place in which the digits are different to order the numbers.

$$2.1\overset{\downarrow}{0}7 \qquad 2.1\overset{\downarrow}{3} \qquad 2.1\overset{\downarrow}{8}$$

DIVISION OF DECIMALS

EXAMPLE   Find $0.17\overline{)4.8195}$

Solution:

$$
\begin{array}{r}
28.35 \\
0.17{\overline{\smash{\big)}\,4.8195}} \\
\underline{34} \\
141 \\
\underline{136} \\
59 \\
\underline{51} \\
85 \\
\underline{85} \\
0
\end{array}
$$

MORE DIVISION OF DECIMALS

EXAMPLE   Find $6.8\overline{)8772}$

Solution:

$$
\begin{array}{r}
1290 \\
6.8{\overline{\smash{\big)}\,8772.0}} \\
\underline{68} \\
197 \\
\underline{136} \\
612 \\
\underline{612} \\
0
\end{array}
$$

USING DECIMALS IN A PAYROLL

EXAMPLE 1 Complete the payroll account below.

Name	Hours Worked	Hourly Rate	Gross Pay (hours × rate)	Deductions	Net Pay (Gross − Deductions)
R. Hu	37	$6.50		$21.00	

Solution: hours × rate = gross pay

37 × 6.50 = 240.50

Gross pay is $240.50.

gross pay − deductions = net pay

240.50 − 21.00 = 219.50

Net pay is $219.50.

EXAMPLE 2 Six employees worked 40, 40, 22, 33, 37, and 38 hours in a week. What was the average number of hours worked?

Solution: Add the number of hours.

40 + 40 + 22 + 33 + 37 + 38 = 210

Divide by the number of employees.

$$\overset{35}{6)\overline{210}}$$

The average number of hours worked was 35.

USING DECIMALS IN MILEAGE

EXAMPLE 1 A car has a 16 gallon gas tank. It gets 28 miles per gallon. How far can it be driven on 4.7 full tanks?

Solution: capacity of gas tank × number of full tanks × miles per gallon = distance

16 × 4.7 × 28 = 2105.6

The car can be driven 2105.6 miles.

EXAMPLE 2 A car averages 33.25 miles per gallon. How much gas is needed to travel 1330 miles?

Solution: number of miles ÷ miles per gallon = gallons of gas needed

1330 ÷ 33.25 = 40

40 gallons of gas are needed.

STUDENT RECORD SHEET

Worksheets Completed

Page Number

1	3	5	7	9
2	4	6	8	10
11	13	15	17	19
12	14	16	18	20
21	23	25	27	29
22	24	26	28	30
31	33	35	37	39
32	34	36	38	40
41	43	45	47	49
42	44	46	48	50
51	53	55	57	59
52	54	56	58	60
61	63	65	67	69
62	64	66	68	70
71	73	75	77	79
72	74	76	78	80

Daily Quiz Grades

No.	Score

Check List Skill Mastered

Date

☐ graphing fractions _____

☐ addition _____

☐ subtraction _____

☐ place value _____

☐ writing decimals _____

☐ rounding to tens, ones, and hundreds _____

☐ rounding to tenths and hundredths _____

☐ multiplication and division _____

☐ comparing fractions and decimals _____

☐ ordering decimals _____

☐ division _____

☐ using decimals in a payroll _____

☐ using decimals in mileage _____

Notes

_____

_____

_____

_____

_____

_____

_____

Graphing fractions

Name _____

Date _____

Graph each fraction.

1. $\frac{2}{3}, \frac{1}{7}, \frac{6}{7}, \frac{2}{11}, \frac{9}{11}$

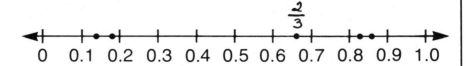

2. $\frac{2}{5}, \frac{1}{6}, \frac{5}{8}, \frac{4}{9}, \frac{4}{11}$

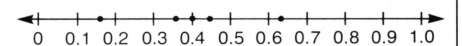

3. $\frac{1}{2}, \frac{4}{5}, \frac{5}{7}, \frac{3}{11}, \frac{8}{11}$

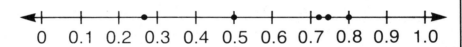

4. $\frac{1}{3}, \frac{3}{4}, \frac{5}{6}, \frac{2}{9}, \frac{5}{11}$

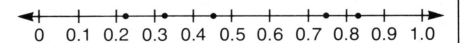

5. $\frac{1}{5}, \frac{2}{7}, \frac{1}{8}, \frac{10}{11}, \frac{7}{12}$

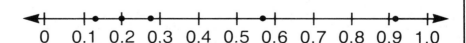

More graphing fractions

Name _____

Date _____

Graph each fraction.

1. $\frac{4}{7}, \frac{3}{8}, \frac{5}{9}, \frac{6}{11}, \frac{1}{13}$

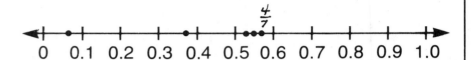

2. $\frac{1}{4}, \frac{1}{9}, \frac{7}{9}, \frac{7}{11}, \frac{2}{13}$

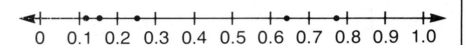

3. $\frac{3}{5}, \frac{3}{7}, \frac{1}{12}, \frac{1}{11}, \frac{5}{16}$

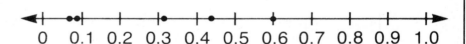

4. $\frac{7}{8}, \frac{8}{9}, \frac{5}{12}, \frac{3}{13}, \frac{7}{16}$

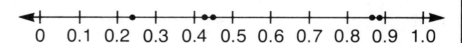

5. $\frac{7}{8}, \frac{4}{9}, \frac{9}{16}, \frac{3}{20}, \frac{8}{13}$

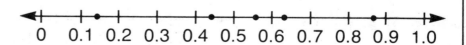

Graphing fractions

Name _____

Date _____

Graph each fraction.

1. $\frac{1}{5}$, $\frac{5}{6}$, $\frac{1}{8}$, $\frac{5}{9}$, $\frac{2}{13}$

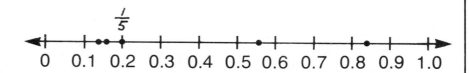

2. $\frac{3}{4}$, $\frac{1}{7}$, $\frac{6}{7}$, $\frac{3}{10}$, $\frac{7}{20}$

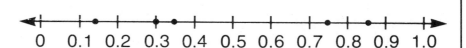

3. $\frac{1}{2}$, $\frac{1}{4}$, $\frac{5}{7}$, $\frac{4}{9}$, $\frac{5}{16}$

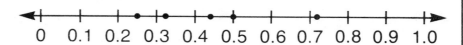

4. $\frac{2}{5}$, $\frac{4}{5}$, $\frac{7}{9}$, $\frac{7}{11}$, $\frac{9}{16}$

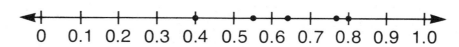

5. $\frac{2}{3}$, $\frac{1}{6}$, $\frac{4}{7}$, $\frac{7}{10}$, $\frac{11}{12}$

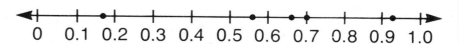

3

More graphing fractions

Name _____

Date _____

Graph each fraction.

1. $\frac{3}{7}$, $\frac{3}{8}$, $\frac{2}{9}$, $\frac{4}{11}$, $\frac{7}{13}$

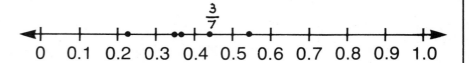

2. $\frac{1}{3}$, $\frac{5}{8}$, $\frac{8}{9}$, $\frac{2}{11}$, $\frac{5}{12}$

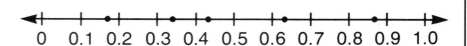

3. $\frac{2}{7}$, $\frac{1}{9}$, $\frac{1}{11}$, $\frac{6}{25}$, $\frac{9}{20}$

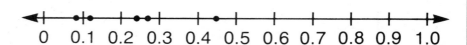

4. $\frac{3}{5}$, $\frac{7}{8}$, $\frac{7}{20}$, $\frac{7}{16}$, $\frac{1}{11}$

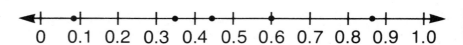

5. $\frac{7}{8}$, $\frac{7}{9}$, $\frac{5}{13}$, $\frac{8}{11}$, $\frac{6}{7}$

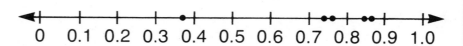

Graphing fractions   Name _____
                                      Date _____

Graph each fraction.

1. $\frac{2}{5}$, $\frac{5}{6}$, $\frac{6}{7}$, $\frac{3}{10}$, $\frac{5}{16}$

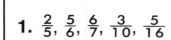
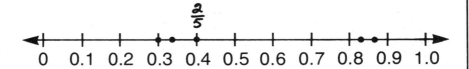

2. $\frac{1}{2}$, $\frac{3}{5}$, $\frac{1}{7}$, $\frac{7}{9}$, $\frac{7}{12}$

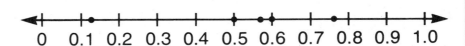

3. $\frac{1}{5}$, $\frac{1}{8}$, $\frac{5}{9}$, $\frac{2}{11}$, $\frac{7}{20}$

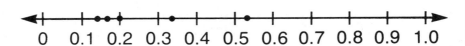

4. $\frac{1}{6}$, $\frac{2}{7}$, $\frac{3}{8}$, $\frac{1}{10}$, $\frac{6}{25}$

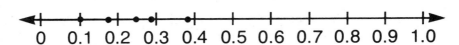

5. $\frac{4}{5}$, $\frac{5}{8}$, $\frac{8}{9}$, $\frac{5}{12}$, $\frac{9}{20}$

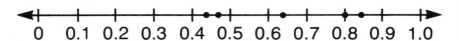

More graphing fractions

Name _____

Date _____

Graph each fraction.

1. $\frac{1}{2}$, $\frac{3}{7}$, $\frac{1}{9}$, $\frac{9}{11}$, $\frac{7}{15}$

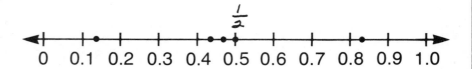

2. $\frac{4}{7}$, $\frac{4}{9}$, $\frac{7}{10}$, $\frac{7}{18}$, $\frac{11}{20}$

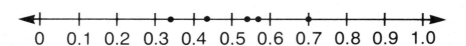

3. $\frac{2}{3}$, $\frac{5}{7}$, $\frac{3}{11}$, $\frac{13}{16}$, $\frac{5}{13}$

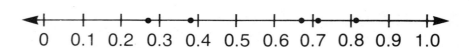

4. $\frac{1}{4}$, $\frac{2}{9}$, $\frac{1}{11}$, $\frac{11}{30}$, $\frac{11}{15}$

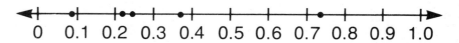

5. $\frac{3}{5}$, $\frac{7}{20}$, $\frac{11}{15}$, $\frac{7}{12}$, $\frac{5}{17}$

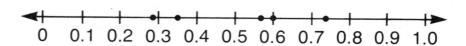

6

Graphing fractions

Name _____

Date _____

Graph each fraction.

1. $\frac{2}{3}$, $\frac{1}{4}$, $\frac{3}{7}$, $\frac{8}{9}$, $\frac{3}{4}$

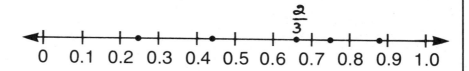

2. $\frac{4}{7}$, $\frac{1}{3}$, $\frac{5}{6}$, $\frac{7}{9}$, $\frac{1}{12}$

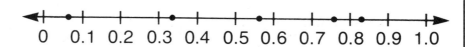

3. $\frac{5}{9}$, $\frac{4}{11}$, $\frac{7}{12}$, $\frac{1}{6}$, $\frac{1}{11}$

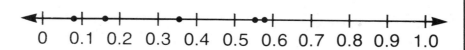

4. $\frac{6}{7}$, $\frac{3}{8}$, $\frac{4}{9}$, $\frac{5}{8}$, $\frac{11}{12}$

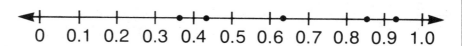

5. $\frac{1}{8}$, $\frac{5}{11}$, $\frac{2}{9}$, $\frac{7}{8}$, $\frac{5}{16}$

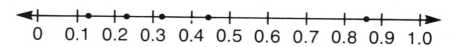

More graphing fractions

Name _____

Date _____

Graph each fraction.

1. $\frac{5}{7}$, $\frac{3}{11}$, $\frac{5}{12}$, $\frac{6}{7}$, $\frac{6}{11}$

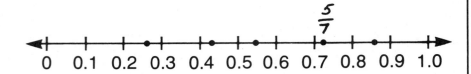

2. $\frac{8}{11}$, $\frac{5}{12}$, $\frac{2}{3}$, $\frac{3}{7}$, $\frac{1}{11}$

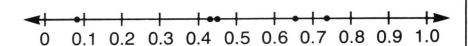

3. $\frac{7}{12}$, $\frac{5}{6}$, $\frac{3}{8}$, $\frac{7}{9}$, $\frac{1}{9}$

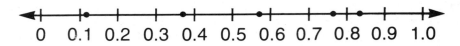

4. $\frac{7}{8}$, $\frac{3}{4}$, $\frac{1}{3}$, $\frac{2}{9}$, $\frac{2}{11}$

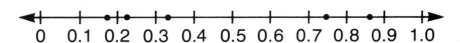

5. $\frac{7}{10}$, $\frac{8}{15}$, $\frac{4}{11}$, $\frac{9}{13}$, $\frac{7}{16}$

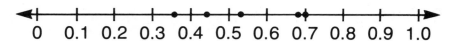

Graphing fractions

Name _____

Date _____

Graph each fraction.

1. $\frac{3}{4}$, $\frac{3}{7}$, $\frac{3}{11}$, $\frac{3}{10}$, $\frac{3}{8}$

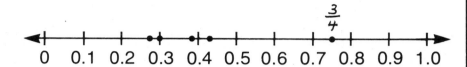

2. $\frac{5}{6}$, $\frac{7}{9}$, $\frac{1}{4}$, $\frac{2}{11}$, $\frac{1}{12}$

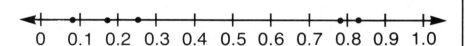

3. $\frac{6}{7}$, $\frac{6}{11}$, $\frac{2}{7}$, $\frac{1}{11}$, $\frac{5}{8}$

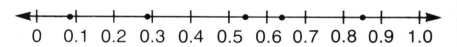

4. $\frac{5}{7}$, $\frac{5}{9}$, $\frac{5}{12}$, $\frac{5}{11}$, $\frac{5}{8}$

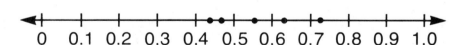

5. $\frac{3}{4}$, $\frac{3}{7}$, $\frac{3}{8}$, $\frac{3}{11}$, $\frac{3}{13}$

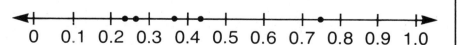

More graphing fractions

Name _____

Date _____

Graph each fraction.

1. $\frac{1}{11}$, $\frac{3}{11}$, $\frac{4}{11}$, $\frac{7}{11}$, $\frac{9}{11}$

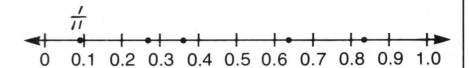

2. $\frac{7}{8}$, $\frac{1}{4}$, $\frac{1}{6}$, $\frac{7}{12}$, $\frac{11}{12}$

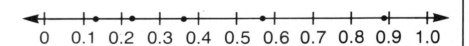

3. $\frac{1}{8}$, $\frac{4}{7}$, $\frac{2}{9}$, $\frac{8}{9}$, $\frac{4}{11}$

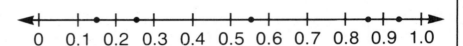

4. $\frac{8}{11}$, $\frac{4}{7}$, $\frac{5}{8}$, $\frac{4}{9}$, $\frac{2}{11}$

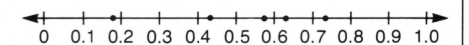

5. $\frac{7}{10}$, $\frac{11}{12}$, $\frac{5}{9}$, $\frac{6}{7}$, $\frac{3}{8}$

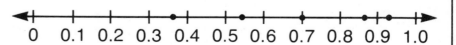

Addition of decimals

Find each sum.

1. 29.6 133.25 16.7 +104.06 *283.61*	**6.** 86.53 194.2 16.78 +309.61
2. 36.23 104.7 9.91 + 16.42	**7.** 802.53 61.92 305.5 + 16.24
3. 17.5 812.63 91.71 +106.05	**8.** 91.26 837.59 16.2 +549.67
4. 36.24 3.756 81.24 +963.5	**9.** Add 38.632, 953.21, 637.45 and 93.127.
5. 92.8 116.43 87.29 +306.5	**10.** Add 26.943, 81.61, 59.32, and 12.

Subtraction of decimals

Name _____

Date _____

Find each difference.

1.	379.26 − 43.82 *335.44*	**6.**	398.21 −186.76
2.	567.59 −108.20	**7.**	969.48 −302.07
3.	137.61 − 48.65	**8.**	304.01 − 69.40
4.	293.87 −182.35	**9.**	Subtract 13.629 from 20.
5.	674.32 − 82.65	**10.**	Subtract 13 from 81.395.

Addition of decimals

Name _____

Date _____

Find each sum.

1. 152.7 63.81 14.006 +106.34 _____ *336.856*	**6.** 103.75 36.821 584.296 + 3.74 _____
2. 83.69 128.413 81.072 + 65.43 _____	**7.** 831.67 69.584 95.26 +384.513 _____
3. 92.31 146.653 94.06 +382.957 _____	**8.** 962.58 3954.1 827.43 +431.7 _____
4. 26.37 592.81 678.05 + 34.87 _____	**9.** Add 757.38, 203.754, and 969.53.
5. 35.96 814.612 39.256 +346.5 _____	**10.** Add 28.672, 341.86, and 105.237.

Subtraction of decimals

Date _____

Find each difference.

1.	539.84 − 67.531 *472.309*	**6.**	342.96 −153.85
2.	926.53 −835.612	**7.**	678.36 −257.83
3.	651.43 − 29.675	**8.**	777.04 − 92.67
4.	831.57 −394.6	**9.**	Subtract 55.367 from 100.
5.	297.38 − 39.76	**10.**	Subtract 302.673 from 690.54.

Addition of decimals

Name _____

Date _____

Find each sum.

1. $\begin{array}{r} 17.62 \\ 319.45 \\ 62.7 \\ +\ \ 83.946 \\ \hline 483.716 \end{array}$	**6.** $\begin{array}{r} 68.73 \\ 284.67 \\ 69.352 \\ +\ \ \ 3.967 \\ \hline \end{array}$
2. $\begin{array}{r} 341.08 \\ 69.343 \\ 205.91 \\ +\ \ \ 3.967 \\ \hline \end{array}$	**7.** $\begin{array}{r} 83.94 \\ 694.536 \\ 91.04 \\ +866.75 \\ \hline \end{array}$
3. $\begin{array}{r} 96.835 \\ 207.31 \\ 83.2 \\ +464.73 \\ \hline \end{array}$	**8.** $\begin{array}{r} 900.7 \\ 63.42 \\ 437.85 \\ +\ \ 67.43 \\ \hline \end{array}$
4. $\begin{array}{r} 82.95 \\ 340.82 \\ 437.5 \\ +\ \ 61.24 \\ \hline \end{array}$	**9.** Add 6.52, 395.1 and 73.41.
5. $\begin{array}{r} 305.61 \\ 82.09 \\ 43.674 \\ +800.9 \\ \hline \end{array}$	**10.** Add 432.14, 16.552, and 804.7.

Name _____

Date _____

Find each difference.

1.	29.634 − 6.07 **23.564**	**6.**	827.56 −343.927
2.	365.479 −128.53	**7.**	609.052 −158.73
3.	694.32 −681.592	**8.**	13.96 − 8.58
4.	936.84 −274.61	**9.**	Subtract 13.796 from 283.95.
5.	3007.92 −1564.73	**10.**	Subtract 28.583 from 83.7624.

Addition of decimals

Find each sum.

1.
```
    63.174
   801.35
    62.905
 +   3.83
  931.259
```

2.
```
    93.756
   831.27
    76.115
 + 392.76
```

3.
```
    86.154
   354.82
    76.917
 + 832.35
```

4.
```
   617.32
    49.76
    32.34
 +  86.597
```

5.
```
    95.304
   366.27
    39.476
 + 404.72
```

6.
```
   962.54
    36.833
    75.46
 + 407.26
```

7.
```
    97.859
   382.74
   659.421
 +  36.967
```

8.
```
    13.852
   637.42
   904.77
 + 213.63
```

9. Add 13.723, 96.01, 142.5, and 16.

10. Add 16.231, 84.3, and 119.26.

Subtraction of decimals

Name *Tony Saros*

Date _____

Find each difference.

1.	349.45 − 62.374 *287.076*	**6.**	375.881 −262.304
2.	835.692 − 93.81	**7.**	596.007 −359.821
3.	627.431 −459.775	**8.**	617.832 −434.97
4.	639.825 −434.614	**9.**	Subtract 19.539 from 73.84.
5.	234.067 −109.226	**10.**	Subtract 6.327 from 50.

Addition of decimals

Name _____

Date _____

Find each sum.

1.	79.632 813.29 66.361 + 582.956 1542.239	6.	594.33 81.916 437.603 + 99.73
2.	384.747 23.833 600.439 + 183.81	7.	484.735 61.63 372.41 + 68.553
3.	93.6 133.804 58.322 + 867.818	8.	27.383 96.41 392.94 + 46.302
4.	39.678 143.84 93.671 + 888.004	9.	Add 38.822, 75.43, 288.62, and 434.01.
5.	306.25 833.7 924.81 + 936.2	10.	Add 13.52, 614.3, 92.533, and 18.

Subtraction of decimals

Name _____

Date _____

Find each difference.

1.	943.61 − 388.53 ───── *555.08*	**6.**	945.99 − 683.47 ───────
2.	631.4 − 585.63 ───────	**7.**	28.692 − 16.58 ───────
3.	961.235 − 48.62 ───────	**8.**	465.27 − 68.569 ───────
4.	83.005 − 28.607 ───────	**9.**	Subtract 95.322 from 105.
5.	353.82 − 304.916 ───────	**10.**	Subtract 66 from 434.21.

Place value for decimals

Name _____

Date _____

Write each number on the blanks below. The decimal point is in the *and* place.

1. 58.33217   **2.** 5.63629   **3.** 0.939321

4. 83.64892   **5.** 6.93304

	MILLIONS	HUNDRED-THOUSANDS	TEN-THOUSANDS	THOUSANDS	HUNDREDS	TENS	ONES	AND	TENTHS	HUNDREDTHS	THOUSANDTHS	TEN-THOUSANDTHS	HUNDRED-THOUSANDTHS	MILLIONTHS
1.						5	8	.	3	3	2	1	7	
2.														
3.														
4.														
5.														

21

Writing decimals

Name _____

Date _____

Write the numeral for each. Fill in zeros where needed.

1. 9 in the tenths place 7 in the ones place 2 in the hundreds place 5 in the hundredths place _ _ 2 0 7 . 9 5 _ _ _	**6.** Nine hundred thousand fifty-five and seventeen thousandths
2. 7 in the thousands place 4 in the ones place 4 in the tenths place 2 in the ten-thousandths place _ _ _ _ . _ _ _ _	**7.** Fifty-seven and three hundred twenty-five thousandths
3. 2 in the thousandths place 5 in the hundreds place 9 in the thousands place 4 in the hundredths place _ _ _ _ . _ _ _ _	**8.** Fourteen and seven tenths
4. 9 in the tens place 7 in the ten-thousands place 3 in the thousands place 1 in the tenths place _ _ _ _ _ . _ _ _ _ _	**9.** Six million five hundred thousand and seven tenths
5. Twelve thousand ninety-two and fifteen hundredths _ _ _ _ _ . _ _ _ _	**10.** Thirty-five thousand fifty-two and twenty-nine hundredths

Place value for decimals   Name _____

                                  Date _____

Write each number on the blanks below. The decimal point is in the *and* place.

1. 35.736   **2.** 912.403   **3.** 7612.57
4. 96.005   **5.** 352.17

	MILLIONS	HUNDRED-THOUSANDS	TEN-THOUSANDS	THOUSANDS	HUNDREDS	TENS	ONES	AND	TENTHS	HUNDREDTHS	THOUSANDTHS	TEN-THOUSANDTHS	HUNDRED-THOUSANDTHS	MILLIONTHS
1.						3	5	.	7	3	6			
2.														
3.														
4.														
5.														

Writing decimals

Name _____

Date _____

Write the numeral for each. Fill in zeros where needed.

1. 7 in the tenths place
   3 in the hundreds place
   2 in the ones place
   4 in the hundredths place

   _ _ 3 0 2 . 7 4 _ _ _

2. 8 in the ones place
   2 in the thousands place
   5 in the tenths place
   1 in the ten-thousandths place

   _ _ _ _ . _ _ _ _

3. 9 in the thousandths place
   5 in the hundreds place
   7 in the thousands place
   4 in the hundredths place

   _ _ _ _ . _ _ _ _

4. 5 in the tens place
   2 in the ten-thousands place
   1 in the thousands place
   4 in the tenths place

   _ _ _ _ . _ _ _ _

5. Thirty five and four thousandths

   _ _ _ _ . _ _ _ _

6. Twenty-nine thousand four hundred and fifteen hundredths

7. Eight thousand one hundred fourteen and twenty-seven thousandths

8. Nine million four hundred thousand and fifteen hundredths

9. Six thousand four hundred twelve and one hundred twenty-two thousandths

10. Two hundred twelve and eight thousandths

Place value for decimals

Name _____

Date _____

Write each number on the blanks below. The decimal point is in the *and* place.

1. 638.921   **2.** 73.69401   **3.** 5.68442
4. 805.332   **5.** 32.61143

MILLIONS	HUNDRED-THOUSANDS	TEN-THOUSANDS	THOUSANDS	HUNDREDS	TENS	ONES	AND	TENTHS	HUNDREDTHS	THOUSANDTHS	TEN-THOUSANDTHS	HUNDRED-THOUSANDTHS	MILLIONTHS
1.				6	3	8	.	9	2	1			
2.													
3.													
4.													
5.													

Writing decimals

Name _____

Date _____

Write the numeral for each. Fill in zeros where needed.

1. 5 in the tenths place 2 in the hundreds place 8 in the ones place 1 in the hundredths place _ _ 2 0 8 . 5 1 _ _ _	**6.** Twenty-two thousand four hundred and seventeen hundredths
2. 8 in the tenths place 4 in the hundredths place 6 in the ones place 4 in the thousands place _ _ _ _ . _ _ _ _	**7.** Two thousand nine and seven tenths
3. 7 in the thousandths place 3 in the hundreds place 2 in the thousands place 2 in the hundredths place _ _ _ _ . _ _ _ _	**8.** Fifty-two thousandths
4. 1 in the thousandths place 4 in the tenths place 4 in the tens place 5 in the ten-thousands place _ _ _ _ . _ _ _ _	**9.** Nine thousand four hundred ten and seventy-five hundredths
5. Twenty-seven and four hundredths _ _ _ _ . _ _ _ _	**10.** Twelve thousand ten and fifty-nine thousandths

Copyright © 1981 by Dale Seymour Publications.

26

Place value for decimals

Name _____

Date _____

Write each number on the blanks below. The decimal point is in the *and* place.

1. 363.8145   **2.** 912.403   **3.** 7612.57

4. 96.005   **5.** 352.17

	MILLIONS	HUNDRED-THOUSANDS	TEN-THOUSANDS	THOUSANDS	HUNDREDS	TENS	ONES	AND	TENTHS	HUNDREDTHS	THOUSANDTHS	TEN-THOUSANDTHS	HUNDRED-THOUSANDTHS	MILLIONTHS
1.	—	—	—	—	3	6	3	.	8	1	4	5	—	—
2.	—	—	—	—	—	—	—	—	—	—	—	—	—	—
3.	—	—	—	—	—	—	—	—	—	—	—	—	—	—
4.	—	—	—	—	—	—	—	—	—	—	—	—	—	—
5.	—	—	—	—	—	—	—	—	—	—	—	—	—	—

Writing decimals

Name _____

Date _____

Write the numeral for each. Fill in zeros where needed.

1. 6 in the tenths place 5 in the hundreds place 9 in the ones place 5 in the hundredths place _ _ 5 0 9 . 6 5 _ _ _	**6.** Six hundred twenty-five and four hundredths
2. 5 in the ones place 8 in the thousands place 9 in the tenths place 4 in the hundredths place _ _ _ _ _ . _ _ _ _	**7.** Three thousand five and one tenth
3. 2 in the thousandths place 6 in the hundreds place 3 in the thousands place 2 in the hundredths place _ _ _ _ _ . _ _ _ _	**8.** Sixty-two and one hundred thirty-two thousandths
4. 1 in the tens place 6 in the ten-thousands place 2 in the thousandths place 7 in the tenths place _ _ _ _ _ . _ _ _ _	**9.** Five hundred four and three tenths
5. Fifty-three and seventeen hundredths _ _ _ _ _ . _ _ _ _	**10.** Ninety-two thousand and ninety-two thousandths

Copyright © 1981 by Dale Seymour Publications.

28

Place value for decimals

Name _____

Date _____

Write each number on the blanks below. The decimal point is in the *and* place.

1. 432.719   **2.** 68.537   **3.** 534.0071
4. 0.7235   **5.** 9683.21

MILLIONS	HUNDRED-THOUSANDS	TEN-THOUSANDS	THOUSANDS	HUNDREDS	TENS	ONES	AND	TENTHS	HUNDREDTHS	THOUSANDTHS	TEN-THOUSANDTHS	HUNDRED-THOUSANDTHS	MILLIONTHS
1. __	__	__	__	4	3	2	.	7	1	9	__	__	__
2. __	__	__	__	__	__	__	__	__	__	__	__	__	__
3. __	__	__	__	__	__	__	__	__	__	__	__	__	__
4. __	__	__	__	__	__	__	__	__	__	__	__	__	__
5. __	__	__	__	__	__	__	__	__	__	__	__	__	__

Name _____

Date _____

Write the numeral for each. Fill in zeros where needed.

1. 5 in the ones place 4 in the thousandths place 3 in the hundreds place 7 in the tenths place _ _ 3 0 5 . 7 0 4 _ _	**6.** Seventeen and four thousandths
2. 6 in the thousands place 8 in the tenths place 3 in the hundredths place 5 in the ones place _ _ _ _ . _ _ _ _	**7.** Seven hundred thirteen and forty-five hundredths
3. 7 in the hundreds place 6 in the thousandths place 8 in the ones place 9 in the ten-thousandths place _ _ _ _ _ . _ _ _ _	**8.** Sixty-two and one hundred thirty-two thousandths
4. 1 in the thousands place 3 in the tenths place 2 in the thousandths place 4 in the hundreds place _ _ _ _ _ . _ _ _ _	**9.** Four hundred thousand and four hundred-thousandths
5. Ten and five hundredths _ _ _ _ _ . _ _ _ _	**10.** Eight hundred twenty-six thousandths

Rounding to the ones, tens, and hundreds places

Name _____

Date _____

Round each number to the ones, tens, and hundreds places.

	ones	tens	hundreds
1. 735.627	736	740	700
2. 6094.357			
3. 439.261			
4. 7643.247			
5. 65.9012			
6. 13.843			
7. 673.106			
8. 571.822			
9. 1306.711			
10. 5363.254			

Copyright © 1981 by Dale Seymour Publications.

31

Rounding to the tenths and hundreths
places

Name _____

Date _____

Round each number to the tenths and hundredths places.

	tenths	hundredths
1. 735.627	735.6	735.63
2. 6094.357		
3. 439.261		
4. 7643.247		
5. 65.9012		
6. 13.843		
7. 673.106		
8. 571.822		
9. 1306.711		
10. 5363.254		

Rounding to the ones, tens, and hundreds
places

Name _____

Date _____

Round each number to the ones, tens, and hundreds places.

	ones	tens	hundreds
1. 2937.658	2938	2940	2900
2. 314.275			
3. 834.813			
4. 29.767			
5. 158.675			
6. 169.132			
7. 377.817			
8. 188.614			
9. 5321.466			
10. 9562.731			

Rounding to the tenths and hundredths places

Name _____

Date _____

Round each number to the tenths and hundredths places.

	tenths	hundredths
1. 2937.658	2937.7	2937.66
2. 314.275		
3. 834.813		
4. 29.767		
5. 158.675		
6. 169.132		
7. 377.817		
8. 188.614		
9. 5321.466		
10. 9562.731		

Rounding to the ones, tens, and hundreds places

Round each number to the ones, tens, and hundreds places.

	ones	tens	hundreds
1. 296.8352	297	300	300
2. 537.1649			
3. 823.572			
4. 2369.634			
5. 538.266			
6. 8359.037			
7. 1621.532			
8. 322.461			
9. 597.802			
10. 763.449			

Rounding to the tenths and hundredths places

Name _____

Date _____

Round each number to the tenths and hundredths places.

	tenths	hundredths
1. 296.8352	296.8	296.84
2. 537.1649		
3. 823.572		
4. 2369.634		
5. 538.266		
6. 8359.037		
7. 1621.532		
8. 322.461		
9. 597.802		
10. 763.449		

Rounding to the ones, tens, and hundreds places   Name _____

Date _____

Round each number to the ones, tens, and hundreds places.

	ones	tens	hundreds
1. 476.319	476	480	500
2. 379.0437			
3. 815.6945			
4. 3214.604			
5. 10,652.389			
6. 583.427			
7. 10,753.923			
8. 836.4814			
9. 297.838			
10. 5673.048			

37

Rounding to the tenths and hundredths
places

Name _____

Date _____

Round each number to the tenths and hundredths places.

	tenths	hundredths
1. 476.319	476.3	476.32
2. 379.0437		
3. 815.6945		
4. 3214.604		
5. 10,652.389		
6. 583.427		
7. 10,753.923		
8. 836.4814		
9. 297.838		
10. 5673.048		

Rounding to the ones, tens, and hundreds
places

Name _____

Date _____

Round each number to the ones, tens, and hundreds places.

	ones	tens	hundreds
1. 117.36201	*117*	*120*	*100*
2. 217.496			
3. 382.671			
4. 912.8526			
5. 136.2127			
6. 9534.486			
7. 136.0953			
8. 106.563			
9. 922.437			
10. 8537.294			

Rounding to the tenths and hundredths
places

Name _____

Date _____

Round each number to the tenths and hundredths places.

	tenths	hundredths
1. 117.36201	117.4	117.36
2. 217.496		
3. 382.671		
4. 912.8526		
5. 136.2127		
6. 9534.486		
7. 136.0953		
8. 106.563		
9. 922.437		
10. 8537.294		

Multiplication and division of decimals   Name _____

                                                 Date _____

Find each product.   Find each quotient.

1. $69.732 \times 10^2 = 6973.2$	**9.** $69.732 \div 100 =$
2. $32.6 \times 10^3 =$	**10.** $32.6 \div 1000 =$
3. $4.397 \times 10 =$	**11.** $4.397 \div 10 =$
4. $63.811 \times 10^2 =$	**12.** $63.811 \div 100 =$
5. $32.905 \times 10^3 =$	**13.** $32.905 \div 1000 =$
6. $\begin{array}{r} 63.912 \\ \times\ \ 0.406 \\ \hline \end{array}$	**14.** $0.83\overline{)2.9133}$
7. $\begin{array}{r} 7.2695 \\ \times\ \ \ \ 82.5 \\ \hline \end{array}$	**15.** $62.7\overline{)283.404}$
8. $\begin{array}{r} 36.4371 \\ \times\ \ \ \ \ 8.22 \\ \hline \end{array}$	**16.** $8.19\overline{)55.1187}$

41

Comparing fractions and decimals

Name _____

Date _____

Use $<$, $>$, or $=$ to complete each of the following.

1. $\frac{3}{5}$ ___=___ 0.6	**6.** $\frac{1}{3}$ _____ 0.3	**11.** $\frac{3}{4}$ _____ 0.3
2. $\frac{4}{9}$ _____ 0.5	**7.** $\frac{1}{7}$ _____ 0.1	**12.** $\frac{3}{4}$ _____ 0.7
3. $\frac{1}{2}$ _____ 0.5	**8.** $\frac{5}{9}$ _____ 0.5	**13.** $\frac{5}{8}$ _____ 0.6
4. $\frac{2}{3}$ _____ 0.6	**9.** $\frac{3}{10}$ _____ 0.3	**14.** $\frac{7}{9}$ _____ 0.8
5. $\frac{3}{8}$ _____ 0.3	**10.** $\frac{4}{5}$ _____ 0.9	**15.** $\frac{6}{7}$ _____ 0.9

Multiplication and division of decimals

Name _____

Date _____

Find each product.

1. $53.624 \times 10^2 = $ *5362.4*
2. $1.7312 \times 10^3 = $
3. $0.814 \times 10 = $
4. $3.8114 \times 10^2 = $
5. $63.52 \times 10^3 = $
6. $\begin{array}{r} 6.371 \\ \times \quad 8.7 \\ \hline \end{array}$
7. $\begin{array}{r} 4.309 \\ \times \quad 8.06 \\ \hline \end{array}$
8. $\begin{array}{r} 7.631 \\ \times \quad 3.14 \\ \hline \end{array}$

Find each quotient.

9. $53.624 \div 100 = $
10. $1.7312 \div 1000 = $
11. $0.814 \div 10 = $
12. $3.8114 \div 10^2 = $
13. $63.52 \div 10^3 = $
14. $6.38 \overline{)2.7434}$
15. $0.71 \overline{)3.9902}$
16. $23.6 \overline{)1980.04}$

Comparing fractions and decimals

Name _____

Date _____

Use $<$, $>$, or $=$ to complete each of the following.

1. $\frac{2}{5}$ __=__ 0.4	**6.** $\frac{1}{2}$ _____ 0.5	**11.** $\frac{2}{3}$ _____ 0.5
2. $\frac{1}{3}$ _____ 0.3	**7.** $\frac{4}{5}$ _____ 0.8	**12.** $\frac{5}{7}$ _____ 0.3
3. $\frac{2}{7}$ _____ 0.4	**8.** $\frac{3}{8}$ _____ 0.2	**13.** $\frac{3}{4}$ _____ 0.9
4. $\frac{5}{9}$ _____ 0.6	**9.** $\frac{3}{10}$ _____ 0.3	**14.** $\frac{8}{10}$ _____ 0.7
5. $\frac{4}{7}$ _____ 0.4	**10.** $\frac{5}{11}$ _____ 0.4	**15.** $\frac{3}{7}$ _____ 0.4

Multiplication and division of decimals

Name _____

Date _____

Find each product.

1. $86.219 \times 10 = 862.19$

2. $3.65 \times 10^2 =$

3. $453.2 \times 10^3 =$

4. $0.415 \times 10^2 =$

5. $7.314 \times 10^3 =$

6.
$$\begin{array}{r} 9.104 \\ \times\quad 6.3 \\ \hline \end{array}$$

7.
$$\begin{array}{r} 26.22 \\ \times\quad 8.01 \\ \hline \end{array}$$

8.
$$\begin{array}{r} 53.43 \\ \times\quad 9.17 \\ \hline \end{array}$$

Find each quotient.

9. $86.219 \div 10 =$

10. $3.65 \div 100 =$

11. $453.2 \div 1000 =$

12. $0.415 \div 10^2 =$

13. $7.314 \div 10^3 =$

14. $4.61\overline{)24.1103}$

15. $0.77\overline{)24.178}$

16. $42.9\overline{)160.446}$

Comparing fractions and decimals

Use $<$, $>$, or $=$ to complete each of the following.

1. $\frac{1}{8}$ __<__ 0.2	**6.** $\frac{2}{5}$ ____ 0.4	**11.** $\frac{3}{4}$ ____ 0.8
2. $\frac{3}{4}$ ____ 0.6	**7.** $\frac{3}{12}$ ____ 0.1	**12.** $\frac{3}{7}$ ____ 0.4
3. $\frac{5}{7}$ ____ 0.8	**8.** $\frac{7}{10}$ ____ 0.7	**13.** $\frac{4}{5}$ ____ 0.8
4. $\frac{1}{2}$ ____ 0.5	**9.** $\frac{6}{11}$ ____ 0.8	**14.** $\frac{4}{9}$ ____ 0.5
5. $\frac{2}{5}$ ____ 0.2	**10.** $\frac{5}{6}$ ____ 0.8	**15.** $\frac{1}{4}$ ____ 0.2

46

Multiplication and division of decimals

Name _____

Date _____

Find each product.

Find each quotient.

| **1.** $96.321 \times 10 = 963.21$ |
| **2.** $4.365 \times 10^2 =$ |
| **3.** $829.1 \times 10^3 =$ |
| **4.** $0.54 \times 10^2 =$ |
| **5.** $3.672 \times 10^3 =$ |
| **6.** $\begin{array}{r} 8.503 \\ \times\ \ \ 2.6 \\ \hline \end{array}$ |
| **7.** $\begin{array}{r} 1.333 \\ \times\ \ \ 7.9 \\ \hline \end{array}$ |
| **8.** $\begin{array}{r} 39.104 \\ \times\ \ \ 7.32 \\ \hline \end{array}$ |

| **9.** $96.321 \div 10 =$ |
| **10.** $4.365 \div 100 =$ |
| **11.** $829.1 \div 1000 =$ |
| **12.** $0.54 \div 10^2 =$ |
| **13.** $3.672 \div 10^3 =$ |
| **14.** $0.92 \overline{)\,3.4592}$ |
| **15.** $8.14 \overline{)\,539.682}$ |
| **16.** $92.7 \overline{)\,773.118}$ |

47

Comparing fractions and decimals

Use $<$, $>$, or $=$ to complete each of the following.

1. $\frac{4}{5}$ __>__ 0.4	**6.** $\frac{5}{7}$ _____ 0.3	**11.** $\frac{3}{5}$ _____ 0.6
2. $\frac{2}{3}$ _____ 0.6	**7.** $\frac{1}{5}$ _____ 0.2	**12.** $\frac{2}{3}$ _____ 0.7
3. $\frac{3}{8}$ _____ 0.3	**8.** $\frac{5}{6}$ _____ 0.8	**13.** $\frac{4}{5}$ _____ 0.9
4. $\frac{9}{10}$ _____ 0.9	**9.** $\frac{5}{9}$ _____ 0.5	**14.** $\frac{4}{7}$ _____ 0.6
5. $\frac{1}{4}$ _____ 0.3	**10.** $\frac{3}{7}$ _____ 0.5	**15.** $\frac{9}{11}$ _____ 0.9

48

Multiplication and division of decimals

Name _____

Date _____

Find each product.

1. $8.617 \times 10^2 = $ *861.7*

2. $931.2 \times 10^3 = $

3. $0.0052 \times 10 = $

4. $83.74 \times 10^2 = $

5. $1.555 \times 10^3 = $

6.
$$\begin{array}{r} 4.375 \\ \times \quad 8.2 \\ \hline \end{array}$$

7.
$$\begin{array}{r} 32.43 \\ \times \quad 9.65 \\ \hline \end{array}$$

8.
$$\begin{array}{r} 1.0043 \\ \times \quad 6.02 \\ \hline \end{array}$$

Find each quotient.

9. $8.617 \div 100 = $

10. $931.2 \div 1000 = $

11. $0.0052 \div 10 = $

12. $83.74 \div 10^2 = $

13. $1.555 \div 10^3 = $

14. $8.75\overline{)84.6125}$

15. $0.87\overline{)29.754}$

16. $0.17\overline{)0.14875}$

49

Comparing fractions and decimals

Name _____

Date _____

Use <, >, or = to complete each of the following.

1. $\frac{3}{5}$ __>__ 0.2	6. $\frac{1}{4}$ _____ 0.1	11. $\frac{5}{8}$ _____ 0.7
2. $\frac{1}{2}$ _____ 0.5	7. $\frac{4}{5}$ _____ 0.8	12. $\frac{4}{7}$ _____ 0.9
3. $\frac{3}{4}$ _____ 0.7	8. $\frac{7}{10}$ _____ 0.7	13. $\frac{3}{8}$ _____ 0.2
4. $\frac{5}{9}$ _____ 0.6	9. $\frac{3}{11}$ _____ 0.2	14. $\frac{3}{7}$ _____ 0.5
5. $\frac{3}{8}$ _____ 0.4	10. $\frac{3}{16}$ _____ 0.1	15. $\frac{4}{9}$ _____ 0.4

Ordering decimals

Write each of the following in order from least to greatest.

1. 0.23	0.45	0.02	<u>0.02</u>	<u>0.23</u>	<u>0.45</u>
2. 0.7	0.9	0.3	_____	_____	_____
3. 0.79	0.7	0.75	_____	_____	_____
4. 0.07	0.77	0.70	_____	_____	_____
5. 5.93	59.7	598	_____	_____	_____
6. 2.02	1.980	2.009	_____	_____	_____
7. 8.5	85	0.85	_____	_____	_____
8. 9.06	0.91	0.905	_____	_____	_____
9. 1.03	0.988	9.88	_____	_____	_____
10. 0.07	0.007	0.0072	_____	_____	_____

Name _____

Date _____

Write each of the following in order from least to greatest.

1. 1001.3	1001.32	1002.2	_1001.3_	_1001.32_	_1002.2_	
2. 3.695	3.09	3.65	_____	_____	_____	
3. 13.009	13.859	13.439	_____	_____	_____	
4. 27.090	27.080	27.070	_____	_____	_____	
5. 47.83	46.99	47.80	_____	_____	_____	
6. 151.089	151.1	151.999	_____	_____	_____	
7. 3.726	8.327	100.2	_____	_____	_____	
8. 12.1	11.239	3.25	_____	_____	_____	
9. 0.03	4.75	12.4	_____	_____	_____	
10. 3.147	3.047	3.247	_____	_____	_____	

Ordering decimals

Write each of the following in order from least to greatest.

1.	0.63	0.9	0.06	*0.06*	*0.63*	*0.9*
2.	0.08	0.82	0.85	_____	_____	_____
3.	0.63	0.6	0.633	_____	_____	_____
4.	0.33	0.5	0.06	_____	_____	_____
5.	1.82	1.63	1.04	_____	_____	_____
6.	9.6	6.9	5.3	_____	_____	_____
7.	12.31	12.32	12.30	_____	_____	_____
8.	13.02	13.002	13.2	_____	_____	_____
9.	3.6	3.66	3.606	_____	_____	_____
10.	14.7	14.69	14.59	_____	_____	_____

More ordering decimals

Name _____

Date _____

Write each of the following in order from least to greatest.

1.	0.003	3.003	2.003	*0.003*	*2.003*	*3.003*
2.	16.7	15.23	16.9	_____	_____	_____
3.	23.03	23.3	23.003	_____	_____	_____
4.	19.4	10.4	13.4	_____	_____	_____
5.	126.7	12.7	1.267	_____	_____	_____
6.	81.00	81.1	81.2	_____	_____	_____
7.	13.6	13.61	13.71	_____	_____	_____
8.	53.3	11.53	15.13	_____	_____	_____
9.	13.62	13.26	12.63	_____	_____	_____
10.	14.5	11.27	8.36	_____	_____	_____

Ordering decimals

Name _____

Date _____

Write each of the following in order from least to greatest.

1.	0.52	0.513	0.5	<u>0.5</u>	<u>0.513</u>	<u>0.52</u>
2.	0.76	0.799	0.724	_____	_____	_____
3.	0.351	0.3	0.364	_____	_____	_____
4.	3.42	3.75	3.795	_____	_____	_____
5.	8.06	7.35	5.32	_____	_____	_____
6.	6.97	12.97	8.97	_____	_____	_____
7.	101.3	10.3	1.013	_____	_____	_____
8.	4.627	4.067	4.007	_____	_____	_____
9.	2.75	2.075	2.0075	_____	_____	_____
10.	3.53	0.353	0.0353	_____	_____	_____

More ordering decimals

Name _____

Date _____

Write each of the following in order from least to greatest.

1. 29.42	294.2	2.942	*2.942*	*29.42*	*294.2*	
2. 359.6	243.7	183.95	_____	_____	_____	
3. 63.7	75.2	42.37	_____	_____	_____	
4. 99.9	909.9	9009.9	_____	_____	_____	
5. 14	2.361	845.67	_____	_____	_____	
6. 655.2	655.25	655.259	_____	_____	_____	
7. 30.6	3.06	0.306	_____	_____	_____	
8. 14.2	99.37	6.532	_____	_____	_____	
9. 7.4	0.74	740	_____	_____	_____	
10. 13.2	27.4	9.32	_____	_____	_____	

56

Ordering decimals

Name _____

Date _____

Write each of the following in order from least to greatest.

1. 0.869	0.8	0.82	<u>0.8</u>	<u>0.82</u>	<u>0.869</u>
2. 5.667	5.666	5.642	_____	_____	_____
3. 10.304	10.403	10.034	_____	_____	_____
4. 9.005	9.055	9.505	_____	_____	_____
5. 0.3	0.32	0.29	_____	_____	_____
6. 5.273	5.27	5.2	_____	_____	_____
7. 8.95	8.095	8.0095	_____	_____	_____
8. 16.43	1.643	164.3	_____	_____	_____
9. 8.233	8.223	8.22	_____	_____	_____
10. 10.5	10.51	10.6	_____	_____	_____

More ordering decimals

Name _____

Date _____

Write each of the following in order from least to greatest.

1.	14.31	31.14	11.34	_11.34_	_14.31_	_31.14_
2.	6.07	6.7	60.7	_____	_____	_____
3.	0.375	375	37.5	_____	_____	_____
4.	15.34	15.43	51.44	_____	_____	_____
5.	9.006	9.6	9.66	_____	_____	_____
6.	101.3	101.42	101.01	_____	_____	_____
7.	43.7	43.07	43.007	_____	_____	_____
8.	19.1	19.2	19.0	_____	_____	_____
9.	154.62	154.43	154.66	_____	_____	_____
10.	13.4	13.44	13.444	_____	_____	_____

Ordering decimals

Write each of the following in order from least to greatest.

1.	0.4321	0.3241	0.1234	_0.1234_	_0.3241_	_0.4321_
2.	5.6952	5.6	5.9	_____	_____	_____
3.	0.8271	8.827	0.0827	_____	_____	_____
4.	15.34	15.442	15.265	_____	_____	_____
5.	16.3	16.35	16.4	_____	_____	_____
6.	3.543	3.54	3.5	_____	_____	_____
7.	12.9	8.361	3.531	_____	_____	_____
8.	5.55	5.45	5.65	_____	_____	_____
9.	10.85	10.9	10.327	_____	_____	_____
10.	38.56	3.856	385.6	_____	_____	_____

More ordering decimals

Name _____

Date _____

Write each of the following in order from least to greatest.

1.	22.2	222	2.22	_2.22_	_22.2_	_222_
2.	3.07	3.7	3.007	_____	_____	_____
3.	93.2	74.567	8.2	_____	_____	_____
4.	191.65	303	141.67	_____	_____	_____
5.	18.75	18.57	18	_____	_____	_____
6.	0.737	0.7	0.3	_____	_____	_____
7.	49.56	49.1	49.32	_____	_____	_____
8.	6.43	6.3	6.343	_____	_____	_____
9.	81.004	81.4	81.04	_____	_____	_____
10.	100.7	107.1	103.2	_____	_____	_____

Division of decimals

Find each quotient.

1. $4.6\overline{)199.042}$ quotient: 43.27	**5.** $3.1\overline{)19.5579}$
2. $0.22\overline{)1301.52}$	**6.** $0.44\overline{)3.18296}$
3. $1.6\overline{)149.44}$	**7.** $0.48\overline{)246.864}$
4. $0.027\overline{)239247}$	**8.** $0.37\overline{)222185}$

61

More division of decimals

Name _____

Date _____

Find each quotient.

1. $2.4\overline{)197.208}$ quotient: 82.17	**5.** $0.23\overline{)150.3533}$
2. $0.56\overline{)206752}$	**6.** $0.41\overline{)3.515012}$
3. $4.5\overline{)37.053}$	**7.** $0.092\overline{)6371}$
4. $0.18\overline{)5.40378}$	

Division of decimals

Find each quotient.

1. $0.52\overline{)14.04}$ with 27 written above	**5.** $7.1\overline{)0.4615}$
2. $0.21\overline{)6.93}$	**6.** $9.3\overline{)7.626}$
3. $0.43\overline{)27.95}$	**7.** $0.062\overline{)167.4}$
4. $0.032\overline{)0.2944}$	**8.** $0.017\overline{)7.31}$

63

More division of decimals

Name _____

Date _____

Find each quotient.

1. $3.5\overline{)0.630}$ _0.18_	**5.** $0.021\overline{)756}$
2. $0.43\overline{)1.118}$	**6.** $4.8\overline{)254.4}$
3. $0.17\overline{)156.4}$	**7.** $0.22\overline{)5.94}$
4. $1.4\overline{)11.90}$	

Division of decimals

Find each quotient.

1. $0.23\overline{)41.4}$ with quotient 180 written above	**5.** $0.92\overline{)1.472}$
2. $1.7\overline{)7.82}$	**6.** $0.47\overline{)24.91}$
3. $0.032\overline{)9.28}$	**7.** $0.032\overline{)8.32}$
4. $4.1\overline{)29.93}$	**8.** $5.8\overline{)27.26}$

65

Find each quotient.

1. $0.45 \overline{)27.45}$ (with 67 written above)

2. $5.2 \overline{)14.56}$

3. $4.1 \overline{)27.47}$

4. $0.27 \overline{)248.4}$

5. $0.52 \overline{)161.2}$

6. $1.7 \overline{)0.731}$

7. $0.62 \overline{)36.58}$

Division of decimals

Name _____

Date _____

Find each quotient.

1. $1.5 \overline{)5.55}$ with quotient 3.7	**5.** $6.2 \overline{)15.50}$
2. $0.26 \overline{)11.18}$	**6.** $2.7 \overline{)17.01}$
3. $0.52 \overline{)45.24}$	**7.** $0.32 \overline{)25.92}$
4. $0.14 \overline{)4.62}$	**8.** $4.1 \overline{)2.296}$

Name _____

Date _____

Find each quotient.

1. $0.019\overline{)9.88}$ $\overset{520}{}$	**5.** $0.049\overline{)78.4}$
2. $0.024\overline{)4.08}$	**6.** $2.2\overline{)39.6}$
3. $0.43\overline{)15.48}$	**7.** $0.47\overline{)1.081}$
4. $0.51\overline{)24.48}$	

Division of decimals

Find each quotient.

1. 8.3$\overline{)3.901}$ $\overset{0.47}{}$	**5.** 0.23$\overline{)17.25}$
2. 2.4$\overline{)1.656}$	**6.** 0.42$\overline{)3.612}$
3. 0.73$\overline{)59.86}$	**7.** 7.5$\overline{)10.50}$
4. 2.9$\overline{)13.34}$	**8.** 0.082$\overline{)56.58}$

More division of decimals

Name _____

Date _____

Find each quotient.

1. $0.55\overline{)2.365}$ with answer 4.3 above	**5.** $3.3\overline{)52.8}$
2. $0.28\overline{)1.876}$	**6.** $0.051\overline{)132.6}$
3. $0.25\overline{)852.5}$	**7.** $0.44\overline{)212.08}$
4. $0.038\overline{)15.314}$	

Using decimals in a payroll   Name _____

                                               Date _____

Complete the gross pay column, the net pay column, and the totals.

PAYROLL ACCOUNT					
Name	Hours Worked	Hourly Rate	Gross Pay (Hours X Rate)	Deductions	Net Pay (Gross − Deductions)
1. K. Hirst	40	$5.25	$210.00	$15.50	$194.50
2. N. Rivas	40	$4.40		$13.75	
3. M. Arndt	23	$2.11		$ 6.30	
4. M. Torres	31	$2.40		$ 6.55	
5. S. Hart	40	$3.10		$ 7.48	
6. L. Assam	42	$3.26		$8.15	
7. Totals					

8. What is the average number of hours worked?

9. What is the average hourly rate?

10. What is the average gross pay?

11. What is the average deduction?

12. What is the average net pay?

13. What is the total gross pay minus the total deductions?

14. What is the average gross pay minus the average deductions?

Using decimals in mileage

Name _____

Date _____

A car has a 14 gallon gas tank. It gets 19 miles per gallon. How far can it be driven on each of the following numbers of full tanks?

1. 2 full tanks *532 miles*	**3.** $3\frac{1}{2}$ full tanks
2. 4.2 full tanks	**4.** 6.3 full tanks

A car averages 12.6 miles per gallon. How much gas is needed to travel each of the following numbers of miles?

5. 189 miles
6. 1008 miles
7. 2142 miles

Using decimals in a payroll

Name _____

Date _____

Complete the gross pay column, the net pay column, and the totals.

PAYROLL ACCOUNT					
Name	Hours Worked	Hourly Rate	Gross Pay (Hours X Rate)	Deductions	Net Pay (Gross − Deductions)
1. R. Bade	40	$4.95	$198.00	$12.75	$185.25
2. D. Yang	34	$3.65		$ 8.14	
3. T. Karr	40	$5.50		$16.10	
4. A. Cenci	36	$4.45		$11.15	
5. C. Gibbs	40	$3.65		$ 9.45	
6. Y. Martinez	44	$4.86		$12.13	
7. Totals					

8. What is the average number of hours worked?

12. What is the average net pay?

9. What is the average hourly rate?

13. What is the total gross pay minus the total deductions?

10. What is the average gross pay?

14. What is the average gross pay minus the average deductions?

11. What is the average deduction?

Using decimals in mileage

Name _____
Date _____

A car has a 14 gallon gas tank. It gets 17.3 miles per gallon. How far can it be driven on each of the following numbers of full tanks?

1. 3 full tanks *726.6 miles*	**3.** 5.5 full tanks
2. 10 full tanks	**4.** 12.3 full tanks

A car averages 15.3 miles per gallon. How much gas is needed to travel each of the following numbers of miles?

5. 1683 miles
6. 1836 miles
7. 2754 miles

Using decimals in a payroll

Name _____

Date _____

Complete the gross pay column, the net pay column, and the totals.

PAYROLL ACCOUNT					
Name	Hours Worked	Hourly Rate	Gross Pay (Hours X Rate)	Deductions	Net Pay (Gross − Deductions)
1. S. Newnan	40	$3.75	$150.00	$11.15	$138.85
2. T. Marquez	40	$4.75		$13.75	
3. D. Nelson	25	$2.11		$ 6.75	
4. V. Vicente	38	$3.10		$ 6.92	
5. N. Ho Kwan	37	$5.50		$15.55	
6. K. Stuart	30	$3.65		$ 8.28	
7. Totals					

8. What is the average number of hours worked?

9. What is the average hourly rate?

10. What is the average gross pay?

11. What is the average deduction?

12. What is the average net pay?

13. What is the total gross pay minus the total deductions?

14. What is the average gross pay minus the average deductions?

Using decimals in mileage   Name _____

                              Date _____

A car has a 16 gallon gas tank. It gets 18.2 miles per gallon. How far can it be driven on each of the following numbers of full tanks?

1. 4 full tanks *1164.8 miles*	3. 20 full tanks
2. 7.2 full tanks	4. 16.4 full tanks

A car averages 22.3 miles per gallon. How much gas is needed to travel each of the following numbers of miles?

5. 2453 miles
6. 3122 miles
7. 3345 miles

Using decimals in a payroll   Name _____

                                Date _____

Complete the gross pay column, the net pay column, and the totals.

PAYROLL ACCOUNT					
Name	Hours Worked	Hourly Rate	Gross Pay (Hours X Rate)	Deductions	Net Pay (Gross – Deductions)
1. L. Swan	37	$5.25	$194.25	$13.20	$181.05
2. F. Juarez	38	$4.40		$10.15	
3. W. Hsing	38	$4.75		$10.75	
4. K. Shah	40	$4.45		$11.15	
5. A. Bruno	39	$3.65		$ 7.92	
6. H. Osaki	36	$4.86		$10.91	
7. Totals					

8. What is the average number of hours worked?

9. What is the average hourly rate?

10. What is the average gross pay?

11. What is the average deduction?

12. What is the average net pay?

13. What is the total gross pay minus the total deductions?

14. What is the average gross pay minus the average deductions?

Name _____

Date _____

A car has an 18 gallon gas tank. It gets 14.7 miles per gallon. How far can it be driven on each of the following numbers of full tanks?

1. 3 full tanks *793.8 miles*	**3.** 4.7 full tanks
2. 10 full tanks	**4.** 8.3 full tanks

A car averages 17.4 miles per gallon. How much gas is needed to travel each of the following numbers of miles?

5. 2088 miles
6. 2262 miles
7. 2784 miles

Using decimals in a payroll

Name _____

Date _____

Complete the gross pay column, the net pay column, and the totals.

PAYROLL ACCOUNT					
Name	Hours Worked	Hourly Rate	Gross Pay (Hours X Rate)	Deductions	Net Pay (Gross – Deductions)
1. R. Nguyen	37	$3.10	$114.70	$ 6.75	$107.95
2. S. Wong	39	$3.24		$ 6.95	
3. J. Ramos	32	$2.40		$ 6.65	
4. M. Strand	39	$4.95		$11.25	
5. E. Perez	40	$4.75		$13.12	
6. B. Watts	36	$3.76		$10.42	
7. Totals					

8. What is the average number of hours worked?

9. What is the average hourly rate?

10. What is the average gross pay?

11. What is the average deduction?

12. What is the average net pay?

13. What is the total gross pay minus the total deductions?

14. What is the average gross pay minus the average deductions?

Using decimals in mileage

Name _____

Date _____

A car has a 20 gallon gas tank. It gets 16.4 miles per gallon. How far can it be driven on each of the following numbers of full tanks?

1. 5 full tanks _1640 miles_	**3.** 8.3 full tanks
2. 7.9 full tanks	**4.** 11.4 full tanks

A car averages 21.2 miles per gallon. How much gas is needed to travel each of the following numbers of miles?

5. 2332 miles
6. 2756 miles
7. 3392 miles

ANSWERS

Page 1

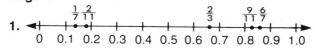

1. (number line with $\frac{1}{7}$, $\frac{2}{11}$, $\frac{2}{3}$, $\frac{9}{11}$, $\frac{6}{7}$)

2. (number line with $\frac{1}{6}$, $\frac{4}{11}$, $\frac{2}{5}$, $\frac{4}{9}$, $\frac{5}{8}$)

3. (number line with $\frac{3}{11}$, $\frac{1}{2}$, $\frac{5}{7}$, $\frac{8}{11}$, $\frac{4}{5}$)

4. (number line with $\frac{2}{9}$, $\frac{1}{3}$, $\frac{5}{11}$, $\frac{3}{4}$, $\frac{5}{6}$)

5. (number line with $\frac{1}{8}$, $\frac{1}{5}$, $\frac{2}{7}$, $\frac{7}{12}$, $\frac{10}{11}$)

Page 2

1. (number line with $\frac{1}{13}$, $\frac{3}{8}$, $\frac{6}{11}$, $\frac{5}{9}$, $\frac{4}{7}$)

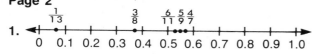

2. (number line with $\frac{1}{9}$, $\frac{2}{13}$, $\frac{1}{4}$, $\frac{7}{11}$, $\frac{7}{9}$)

3. (number line with $\frac{1}{12}$, $\frac{1}{11}$, $\frac{5}{16}$, $\frac{3}{7}$, $\frac{3}{5}$)

4. (number line with $\frac{3}{13}$, $\frac{5}{12}$, $\frac{7}{16}$, $\frac{7}{8}$, $\frac{8}{9}$)

5. (number line with $\frac{3}{20}$, $\frac{4}{9}$, $\frac{9}{16}$, $\frac{8}{13}$, $\frac{7}{8}$)

Page 3

1. (number line with $\frac{1}{8}$, $\frac{2}{13}$, $\frac{1}{5}$, $\frac{5}{9}$, $\frac{5}{6}$)

2. (number line with $\frac{1}{7}$, $\frac{3}{10}$, $\frac{7}{20}$, $\frac{3}{4}$, $\frac{6}{7}$)

3. (number line with $\frac{1}{4}$, $\frac{5}{16}$, $\frac{4}{9}$, $\frac{1}{2}$, $\frac{5}{7}$)

4. (number line with $\frac{2}{5}$, $\frac{9}{16}$, $\frac{7}{11}$, $\frac{7}{9}$, $\frac{4}{5}$)

5. (number line with $\frac{1}{6}$, $\frac{4}{7}$, $\frac{2}{3}$, $\frac{7}{10}$, $\frac{11}{12}$)

Page 4

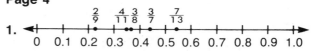

1. (number line with $\frac{2}{9}$, $\frac{4}{11}$, $\frac{3}{8}$, $\frac{3}{7}$, $\frac{7}{13}$)

2. (number line with $\frac{2}{11}$, $\frac{1}{3}$, $\frac{5}{12}$, $\frac{5}{8}$, $\frac{8}{9}$)

3. (number line with $\frac{1}{11}$, $\frac{1}{9}$, $\frac{6}{25}$, $\frac{2}{7}$, $\frac{9}{20}$)

4. (number line with $\frac{1}{11}$, $\frac{7}{20}$, $\frac{7}{16}$, $\frac{3}{5}$, $\frac{7}{8}$)

5. (number line with $\frac{5}{13}$, $\frac{8}{11}$, $\frac{7}{9}$, $\frac{6}{7}$, $\frac{7}{8}$)

Page 5

1. (number line with $\frac{3}{10}$, $\frac{5}{16}$, $\frac{2}{5}$, $\frac{5}{6}$, $\frac{6}{7}$)

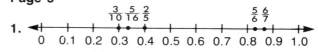

2. (number line with $\frac{1}{7}$, $\frac{1}{2}$, $\frac{7}{12}$, $\frac{3}{5}$, $\frac{7}{9}$)

3. (number line with $\frac{1}{8}$, $\frac{2}{11}$, $\frac{1}{5}$, $\frac{7}{20}$, $\frac{5}{9}$)

4. (number line with $\frac{1}{10}$, $\frac{1}{6}$, $\frac{6}{25}$, $\frac{2}{7}$, $\frac{3}{8}$)

5. (number line with $\frac{5}{12}$, $\frac{9}{20}$, $\frac{5}{8}$, $\frac{4}{5}$, $\frac{8}{9}$)

Page 6

1. (number line with $\frac{1}{9}$, $\frac{3}{7}$, $\frac{7}{15}$, $\frac{1}{2}$, $\frac{9}{11}$)

2. (number line with $\frac{7}{18}$, $\frac{4}{9}$, $\frac{11}{20}$, $\frac{4}{7}$, $\frac{7}{10}$)

3. (number line with $\frac{3}{11}$, $\frac{5}{13}$, $\frac{2}{3}$, $\frac{5}{7}$, $\frac{13}{16}$)

4. (number line with $\frac{1}{11}$, $\frac{2}{9}$, $\frac{1}{4}$, $\frac{11}{30}$, $\frac{11}{15}$)

5. (number line with $\frac{5}{17}$, $\frac{7}{20}$, $\frac{7}{12}$, $\frac{3}{5}$, $\frac{11}{15}$)

Page 7

1. (number line with $\frac{1}{4}$, $\frac{3}{7}$, $\frac{2}{3}$ $\frac{3}{4}$, $\frac{8}{9}$)
2. (number line with $\frac{1}{12}$, $\frac{1}{3}$, $\frac{4}{7}$, $\frac{7}{9}$ $\frac{5}{6}$)
3. (number line with $\frac{1}{11}$ $\frac{1}{6}$, $\frac{4}{11}$, $\frac{5}{9}$ $\frac{7}{12}$)
4. (number line with $\frac{3}{8}$ $\frac{4}{9}$, $\frac{5}{8}$, $\frac{6}{7}$ $\frac{11}{12}$)
5. (number line with $\frac{1}{8}$ $\frac{2}{9}$ $\frac{5}{16}$ $\frac{5}{11}$, $\frac{7}{8}$)

Page 8

1. (number line with $\frac{3}{11}$, $\frac{5}{12}$ $\frac{6}{11}$, $\frac{5}{7}$, $\frac{6}{7}$)
2. (number line with $\frac{1}{11}$, $\frac{5}{12}$ $\frac{3}{7}$, $\frac{2}{3}$ $\frac{8}{11}$)
3. (number line with $\frac{1}{9}$, $\frac{3}{8}$, $\frac{7}{12}$, $\frac{7}{9}$ $\frac{5}{6}$)
4. (number line with $\frac{2}{11}$ $\frac{2}{9}$, $\frac{1}{3}$, $\frac{3}{4}$, $\frac{7}{8}$)
5. (number line with $\frac{4}{11}$ $\frac{7}{16}$ $\frac{8}{15}$, $\frac{9}{13}$ $\frac{7}{10}$)

Page 9

1. (number line with $\frac{3}{11}$ $\frac{3}{10}$ $\frac{3}{8}$ $\frac{3}{7}$, $\frac{3}{4}$)
2. (number line with $\frac{1}{12}$ $\frac{2}{11}$ $\frac{1}{4}$, $\frac{7}{9}$ $\frac{5}{6}$)
3. (number line with $\frac{1}{11}$, $\frac{2}{7}$, $\frac{6}{11}$ $\frac{5}{8}$, $\frac{6}{7}$)
4. (number line with $\frac{5}{12}$ $\frac{5}{11}$ $\frac{5}{9}$ $\frac{5}{8}$ $\frac{5}{7}$)
5. (number line with $\frac{3}{13}$ $\frac{3}{11}$ $\frac{3}{8}$ $\frac{3}{7}$, $\frac{3}{4}$)

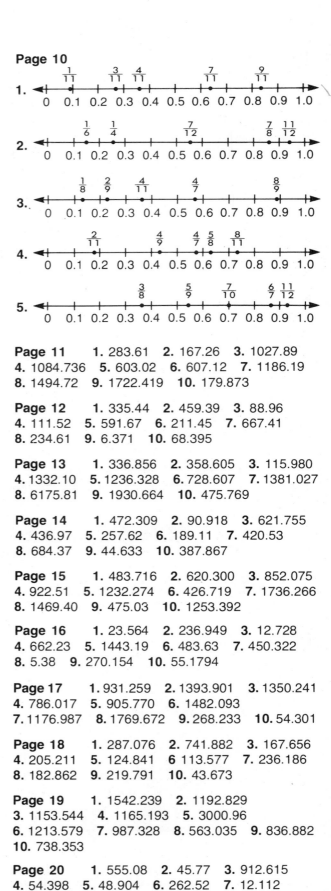

Page 10

1. (number line with $\frac{1}{11}$, $\frac{3}{11}$ $\frac{4}{11}$, $\frac{7}{11}$, $\frac{9}{11}$)
2. (number line with $\frac{1}{6}$ $\frac{1}{4}$, $\frac{7}{12}$, $\frac{7}{8}$ $\frac{11}{12}$)
3. (number line with $\frac{1}{8}$ $\frac{2}{9}$ $\frac{4}{11}$, $\frac{4}{7}$, $\frac{8}{9}$)
4. (number line with $\frac{2}{11}$, $\frac{4}{9}$, $\frac{4}{7}$ $\frac{5}{8}$, $\frac{8}{11}$)
5. (number line with $\frac{3}{8}$, $\frac{5}{9}$, $\frac{7}{10}$, $\frac{6}{7}$ $\frac{11}{12}$)

Page 11   **1.** 283.61   **2.** 167.26   **3.** 1027.89
4. 1084.736   **5.** 603.02   **6.** 607.12   **7.** 1186.19
8. 1494.72   **9.** 1722.419   **10.** 179.873

Page 12   **1.** 335.44   **2.** 459.39   **3.** 88.96
4. 111.52   **5.** 591.67   **6.** 211.45   **7.** 667.41
8. 234.61   **9.** 6.371   **10.** 68.395

Page 13   **1.** 336.856   **2.** 358.605   **3.** 115.980
4. 1332.10   **5.** 1236.328   **6.** 728.607   **7.** 1381.027
8. 6175.81   **9.** 1930.664   **10.** 475.769

Page 14   **1.** 472.309   **2.** 90.918   **3.** 621.755
4. 436.97   **5.** 257.62   **6.** 189.11   **7.** 420.53
8. 684.37   **9.** 44.633   **10.** 387.867

Page 15   **1.** 483.716   **2.** 620.300   **3.** 852.075
4. 922.51   **5.** 1232.274   **6.** 426.719   **7.** 1736.266
8. 1469.40   **9.** 475.03   **10.** 1253.392

Page 16   **1.** 23.564   **2.** 236.949   **3.** 12.728
4. 662.23   **5.** 1443.19   **6.** 483.63   **7.** 450.322
8. 5.38   **9.** 270.154   **10.** 55.1794

Page 17   **1.** 931.259   **2.** 1393.901   **3.** 1350.241
4. 786.017   **5.** 905.770   **6.** 1482.093
7. 1176.987   **8.** 1769.672   **9.** 268.233   **10.** 54.301

Page 18   **1.** 287.076   **2.** 741.882   **3.** 167.656
4. 205.211   **5.** 124.841   **6** 113.577   **7.** 236.186
8. 182.862   **9.** 219.791   **10.** 43.673

Page 19   **1.** 1542.239   **2.** 1192.829
3. 1153.544   **4.** 1165.193   **5.** 3000.96
6. 1213.579   **7.** 987.328   **8.** 563.035   **9.** 836.882
10. 738.353

Page 20   **1.** 555.08   **2.** 45.77   **3.** 912.615
4. 54.398   **5.** 48.904   **6.** 262.52   **7.** 12.112
8. 396.701   **9.** 9.678   **10.** 368.21

Page 21   **1.** 5 tens, 8 ones, and 3 tenths, 3 hundredths, 2 thousandths, 1 ten-thousandth, 7 hundred-thousandths   **2.** 5 ones, and 6 tenths, 3 hundredths, 6 thousandths, 2 ten-thousandths, 9 hundred-thousandths   **3.** 0 ones, and 9 tenths, 3 hundredths, 9 thousandths, 3 ten-thousandths, 2 hundred-thousandths, 1 millionth   **4.** 8 tens, 3 ones, and 6 tenths, 4 hundredths, 8 thousandths, 9 ten-thousandths, 2 hundred-thousandths   **5.** 6 ones, and 9 tenths, 3 hundredths, 3 thousandths, 0 ten-thousandths, 4 hundred-thousandths

Page 22   **1.** 207.95   **2.** 7004.4002   **3.** 9500.042   **4.** 73,090.1   **5.** 12,092.15   **6.** 900,055.017   **7.** 57.325   **8.** 14.7   **9.** 6,500,000.7   **10.** 35,052.29

Page 23   **1.** 3 tens, 5 ones, and 7 tenths, 3 hundredths, 6 thousandths   **2.** 9 hundreds, 1 ten, 2 ones, and 4 tenths, 0 hundredths, 3 thousandths   **3.** 7 thousands, 6 hundreds, 1 ten, 2 ones, and 5 tenths, 7 hundredths   **4.** 9 tens, 6 ones, and 0 tenths, 0 hundredths, 5 thousandths   **5.** 3 hundreds, 5 tens, 2 ones, and 1 tenth, 7 hundredths

Page 24   **1.** 302.74   **2.** 2008.5001   **3.** 7500.039   **4.** 21,050.4   **5.** 35.004   **6.** 29,400.15   **7.** 8114.027   **8.** 9,400,000.15   **9.** 6412.122   **10.** 212.008

Page 25   **1.** 6 hundreds, 3 tens, 8 ones, and 9 tenths, 2 hundredths, 1 thousandth   **2.** 7 tens, 3 ones, and 6 tenths, 9 hundredths, 4 thousandths, 0 ten-thousandths, 1 hundred-thousandth   **3.** 5 ones, and 6 tenths, 8 hundredths, 4 thousandths, 4 ten-thousandths, 2 hundred-thousandths   **4.** 8 hundreds, 0 tens, 5 ones, and 3 tenths, 3 hundredths, 2 thousandths   **5.** 3 tens, 2 ones, and 6 tenths, 1 hundredth, 1 thousandth, 4 ten-thousandths, 3 hundred-thousandths

Page 26   **1.** 208.51   **2.** 4006.84   **3.** 2300.027   **4.** 50,040.401   **5.** 27.04   **6.** 22,400.17   **7.** 2009.7   **8.** 0.052   **9.** 9410.75   **10.** 12,010.059

Page 27   **1.** 3 hundreds, 6 tens, 3 ones, and 8 tenths, 1 hundredth, 4 thousandths, 5 ten-thousandths   **2.** 9 hundreds, 1 ten, 2 ones, and 4 tenths, 0 hundredths, 3 thousandths   **3.** 7 thousands, 6 hundreds, 1 ten, 2 ones, and 5 tenths, 7 hundredths   **4.** 9 tens, 6 ones, and 0 tenths, 0 hundredths, 5 thousandths   **5.** 3 hundreds, 5 tens, 2 ones, and 1 tenth, 7 hundredths

Page 28   **1.** 509.65   **2.** 8005.94   **3.** 3600.022   **4.** 60,010.702   **5.** 53.17   **6.** 625.04   **7.** 3005.1   **8.** 62.132   **9.** 504.3   **10.** 92,000.092

Page 29   **1.** 4 hundreds, 3 tens, 2 ones, and 7 tenths, 1 hundredth, 9 thousandths   **2.** 6 tens, 8 ones, and 5 tenths, 3 hundredths, 7 thousandths

3. 5 hundreds, 3 tens, 4 ones, and 0 tenths, 0 hundredths, 7 thousandths, 1 ten-thousandth   **4.** 0 ones, and 7 tenths, 2 hundredths, 3 thousandths, 5 ten-thousandths   **5.** 9 thousands, 6 hundreds, 8 tens, 3 ones, and 2 tenths, 1 hundredth

Page 30   **1.** 305.704   **2.** 6005.83   **3.** 708.0069   **4.** 1400.302   **5.** 10.05   **6.** 17.004   **7.** 713.45   **8.** 62.132   **9.** 400,000.00004   **10.** 0.826

Page 31   **1.** 736, 740, 700   **2.** 6094, 6090, 6100   **3.** 439, 440, 400   **4.** 7643, 7640, 7600   **5.** 66, 70, 100   **6.** 14, 10, 0   **7.** 673, 670, 700   **8.** 572, 570, 600   **9.** 1307, 1310, 1300   **10.** 5363, 5360 5400

Page 32   **1.** 735.6, 735.63   **2.** 6094.4, 6094.36   **3.** 439.3, 439.26   **4.** 7643.2, 7643.25   **5.** 65.9, 65.90   **6.** 13.8, 13.84   **7.** 673.1, 673.11   **8.** 571.8, 571.82   **9.** 1306.7, 1306.71   **10.** 5363.3, 5363.25

Page 33   **1.** 2938, 2940, 2900   **2.** 314, 310, 300   **3.** 835, 830, 800   **4.** 30, 30, 0   **5.** 159, 160, 200   **6.** 169, 170, 200   **7.** 378, 380, 400   **8.** 189, 190, 200   **9.** 5321, 5320, 5300   **10.** 9563, 9560, 9600

Page 34   **1.** 2937.7, 2937.66   **2.** 314.3, 314.28   **3.** 834.8, 834.81   **4.** 29.8, 29.77   **5.** 158.7, 158.68   **6.** 169.1, 169.13   **7.** 377.8, 377.82   **8.** 188.6, 188.61   **9.** 5321.5, 5321.47   **10.** 9562.7, 9562.73

Page 35   **1.** 297, 300, 300   **2.** 537, 540, 500   **3.** 824, 820, 800   **4.** 2370, 2370, 2400   **5.** 538, 540, 500   **6.** 8359, 8360, 8400   **7.** 1622, 1620, 1600   **8.** 322, 320, 300   **9.** 598, 600, 600   **10.** 763, 760, 800

Page 36   **1.** 296.8, 296.84   **2.** 537.2, 537.16   **3.** 823.6, 823.57   **4.** 2369.6, 2369.63   **5.** 538.3, 538.27   **6.** 8359.0, 8359.04   **7.** 1621.5, 1621.53   **8.** 322.5, 322.46   **9.** 597.8, 597.80   **10.** 763.4, 763.45

Page 37   **1.** 476, 480, 500   **2.** 379, 380, 400   **3.** 816, 820, 800   **4.** 3215, 3210, 3200   **5.** 10,652, 10,650, 10,700   **6.** 583, 580, 600   **7.** 10,754, 10,750, 10,800   **8.** 836, 840, 800   **9.** 298, 300 300   **10.** 5673, 5670, 5700

Page 38   **1.** 476.3, 476.32   **2.** 379.0, 379.04   **3.** 815.7, 815.69   **4.** 3214.6, 3214.60   **5.** 10,652.4, 10,652.39   **6.** 583.4, 583.43   **7.** 10,753.9, 10,753.92   **8.** 836.5, 836.48   **9.** 297.8, 297.84   **10.** 5673.0, 5673.05

Page 39   **1.** 117, 120, 100   **2.** 217, 220, 200   **3.** 383, 380, 400   **4.** 913, 910, 900   **5.** 136, 140 100   **6.** 9534, 9530, 9500   **7.** 136, 140, 100   **8.** 107, 110, 100   **9.** 922, 920, 900   **10.** 8537, 8540, 8500

Page 40   **1.** 117.4, 117.36   **2.** 217.5, 217.50
3. 382.7, 382.67   **4.** 912.9, 912.85   **5.** 136.2,
136.21   **6.** 9534.5, 9534.49   **7.** 136.1, 136.10
8. 106.6, 106.56   **9.** 922.4, 922.44   **10.** 8537.3,
8537.29

Page 41   **1.** 6973.2   **2.** 32,600   **3.** 43.97
4. 6381.1   **5.** 32,905   **6.** 25.948272   **7.** 599.73375
8. 299.512962   **9.** 0.69732   **10.** 0.0326
11. 0.4397   **12.** 0.63811   **13.** 0.032905   **14.** 3.51
15. 4.52   **16.** 6.73

Page 42   **1.** =   **2.** <   **3.** =   **4.** >   **5.** >   **6.** >
7. >   **8.** >   **9.** =   **10.** <   **11.** >   **12.** >   **13.** >
14. <   **15.** <

Page 43   **1.** 5362.4   **2.** 1731.2   **3.** 8.14
4. 381.14   **5.** 63,520   **6.** 55.4277   **7.** 34.73054
8. 23.96134   **9.** 0.53624   **10.** 0.0017312
11. 0.0814   **12.** 0.038114   **13.** 0.06352   **14.** 0.43
15. 5.62   **16.** 83.9

Page 44   **1.** =   **2.** >   **3.** <   **4.** <   **5.** >   **6.** =
7. =   **8.** >   **9.** =   **10.** >   **11.** >   **12.** >   **13.** <
14. >   **15.** >

Page 45   **1.** 862.19   **2.** 365   **3.** 453,200
4. 415   **5.** 7314   **6.** 57.3552   **7.** 210.0222
8. 489.9531   **9.** 8.6219   **10.** 0.0365   **11.** 0.4532
12. 0.00415   **13.** 0.007314   **14.** 5.23   **15.** 31.4
16. 3.74

Page 46   **1.** <   **2.** >   **3.** <   **4.** =   **5.** >   **6.** =
7. >   **8.** =   **9.** <   **10.** >   **11.** <   **12.** >   **13.** =
14. <   **15.** >

Page 47   **1.** 963.21   **2.** 436.5   **3.** 829,100
4. 54   **5.** 3672   **6.** 22.1078   **7.** 10.5307
8. 286.24128   **9.** 9.6321   **10.** 0.04365   **11.** 0.8291
12. 0.0054   **13.** 0.003672   **14.** 3.76   **15.** 66.7
16. 8.34

Page 48   **1.** >   **2.** >   **3.** >   **4.** =   **5.** <   **6.** >
7. =   **8.** >   **9.** >   **10.** <   **11.** =   **12.** <   **13.** <
14. <   **15.** <

Page 49   **1.** 861.7   **2.** 931,200   **3.** 0.052
4. 8374   **5.** 1555   **6.** 35.8750   **7.** 312.9495
8. 6.045886   **9.** 0.08617   **10.** 0.9312   **11.** 0.00052
12. 0.8374   **13.** 0.001555   **14.** 9.67   **15.** 34.2
16. 0.875

Page 50   **1.** >   **2.** =   **3.** >   **4.** <   **5.** <   **6.** >
7. =   **8.** =   **9.** >   **10.** >   **11.** <   **12.** <   **13.** >
14. <   **15.** >

Page 51   **1.** 0.02, 0.23, 0.45   **2.** 0.3, 0.7, 0.9
3. 0.7, 0.75, 0.79   **4.** 0.07, 0.70, 0.77   **5.** 5.93,
59.7, 598   **6.** 1.980, 2.009, 2.02   **7.** 0.85, 8.5, 85
8. 0.905, 0.91, 9.06   **9.** 0.988, 1.03, 9.88
10. 0.007, 0.0072, 0.07

Page 52   **1.** 1001.3, 1001.32, 1002.2   **2.** 3.09,
3.65, 3.695   **3.** 13.009, 13.439, 13.859   **4.** 27.070,
27.080, 27.090   **5.** 46.99, 47.80, 47.83
6. 151.089, 151.1, 151.999   **7.** 3.726, 8.327, 100.2
8. 3.25, 11.239, 12.1   **9.** 0.03, 4.75, 12.4
10. 3.047, 3.147, 3.247

Page 53   **1.** 0.06, 0.63, 0.9   **2.** 0.08, 0.82, 0.85
3. 0.6, 0.63, 0.633   **4.** 0.06, 0.33, 0.5   **5.** 1.04,
1.63, 1.82   **6.** 5.3, 6.9, 9.6   **7.** 12.30, 12.31, 12.32
8. 13.002, 13.02, 13.2   **9.** 3.6, 3.606, 3.66
10. 14.59, 14.69, 14.7

Page 54   **1.** 0.003, 2.003, 3.003   **2.** 15.23, 16.7,
16.9   **3.** 23.003, 23.03, 23.3   **4.** 10.4, 13.4, 19.4
5. 1.267, 12.7, 126.7   **6.** 81.00, 81.1, 81.2   **7.** 13.6,
13.61, 13.71   **8.** 11.53, 15.13, 53.3   **9.** 12.63,
13.26, 13.62   **10.** 8.36, 11.27, 14.5

Page 55   **1.** 0.5, 0.513, 0.52   **2.** 0.724, 0.76,
0.799   **3.** 0.3, 0.351, 0.364   **4.** 3.42, 3.75, 3.795
5. 5.32, 7.35, 8.06   **6.** 6.97, 8.97, 12.97   **7.** 1.013,
10.3, 101.3   **8.** 4.007, 4.067, 4.627   **9.** 2.0075,
2.075, 2.75   **10.** 0.0353, 0.353, 3.53

Page 56   **1.** 2.942, 29.42, 294.2   **2.** 183.95,
243.7, 359 6   **3.** 42.37, 63.7, 75.2   **4.** 99.9, 909.9,
9009.9   **5.** 2.361, 14, 845.67   **6.** 655.2, 655.25,
655.259   **7.** 0.306, 3.06, 30.6   **8.** 6.532, 14.2,
99.37   **9.** 0.74, 7.4, 740   **10.** 9.32, 13.2, 27.4

Page 57   **1.** 0.8, 0.82, 0.869   **2.** 5.642, 5.666,
5.667   **3.** 10.034, 10.304, 10.403   **4.** 9.005,
9.055, 9.505   **5.** 0.29, 0.3, 0.32   **6.** 5.2, 5.27, 5.273
7. 8.0095, 8.095, 8.95   **8.** 1.643, 16.43, 164.3
9. 8.22, 8.223, 8.233   **10.** 10.5, 10.51, 10.6

Page 58   **1.** 11.34, 14.31, 31.14   **2.** 6.07, 6.7,
60.7   **3.** 0.375, 37.5, 375   **4.** 15.34, 15.43, 51.44
5. 9.006, 9.6, 9.66   **6.** 101.01, 101.3, 101.42
7. 43.007, 43.07, 43.7   **8.** 19.0, 19.1, 19.2
9. 154.43, 154.62, 154.66   **10.** 13.4, 13.44, 13.444

Page 59   **1.** 0.1234, 0.3241, 0.4321   **2.** 5.6,
5.6952, 5.9   **3.** 0.0827, 0.8271, 8.827   **4.** 15.265,
15.34, 15.442   **5.** 16.3, 16.35, 16.4   **6.** 3.5, 3.54,
3.543   **7.** 3.531, 8.361, 12.9   **8.** 5.45, 5.55, 5.65
9. 10.327, 10.85, 10.9   **10.** 3.856, 38.56, 385.6

Page 60   **1.** 2.22, 22.2, 222   **2.** 3.007, 3.07,
3.7   **3.** 8.2, 74.567, 93.2   **4.** 141.67, 191.65, 303
5. 18, 18.57, 18.75   **6.** 0.3, 0.7, 0.737   **7.** 49.1,
49.32, 49.56   **8.** 6.3, 6.343, 6.43   **9.** 81.004,
81.04 81.4   **10.** 100.7, 103.2, 107.1

Page 61   **1.** 43.27   **2.** 5916   **3.** 93.4
4. 8,861,000   **5.** 6.309   **6.** 7.234   **7.** 514.3
8. 600,500

Page 62   **1.** 82.17   **2.** 369,200   **3.** 8.234
4. 30.021   **5.** 653.71   **6.** 8.5732   **7.** 69,250

Page 63   **1.** 27   **2.** 33   **3.** 65   **4.** 9.2   **5.** 0.065
6. 0.82   **7.** 2700   **8.** 430

Page 64   **1.** 0.18   **2.** 2.6   **3.** 920   **4.** 8.5
5. 36,000   **6.** 53   **7.** 27

Page 65   **1.** 180   **2.** 4.6   **3.** 290   **4.** 7.3   **5.** 1.6
6. 53   **7.** 260   **8.** 4.7

Page 66   **1.** 61   **2.** 2.8   **3.** 6.7   **4.** 920   **5.** 310
6. 0.43   **7.** 59

Page 67   **1.** 3.7   **2.** 43   **3.** 87   **4.** 33   **5.** 2.5
6. 6.3   **7.** 81   **8.** 0.56

Page 68   **1.** 520   **2.** 170   **3.** 36   **4.** 48   **5.** 1600
6. 18   **7.** 2.3

Page 69   **1.** 0.47   **2.** 0.69   **3.** 82   **4.** 4.6   **5.** 75
6. 8.6   **7.** 1.4   **8.** 690

Page 70.   **1.** 4.3   **2.** 6.7   **3.** 3410   **4.** 403
5. 160   **6.** 2600   **7.** 482

Page 71   **1.** $210.00, $194.50   **2.** $176.00,
$162.25   **3.** $48.53, $42.23   **4.** $74.40, $67.85
5. $124.00, $116.52   **6.** $136.92, $128.77
7. 216, $20.52, $769.85, $57.73, $712.12   **8.** 36
9. $3.42   **10.** $128.31   **11.** $9.62   **12.** $118.69
13. $712.12   **14.** $118.69

Page 72   **1.** 532 miles   **2.** 1117.2 miles
3. 931 miles   **4.** 1675.8 miles   **5.** 15 gallons
6. 80 gallons   **7.** 170 gallons

Page 73   **1.** $198.00, $185.25   **2.** $124.10,
$115.96   **3.** $220.00, $203.90   **4.** $160.20,
$149.05   **5.** $146.00, $136.55   **6.** $213.84,
$201.71   **7.** 234, $27.06, $1062.14, $69.72,
$992.42   **8.** 39   **9.** $4.51   **10.** $177.02
11. $11.62   **12.** $165.40   **13.** $992.42
14. $165.40

Page 74   **1.** 726.6 miles   **2.** 2422 miles
3. 1332.1 miles   **4.** 2979.06 miles   **5.** 110 gallons
6. 120 gallons   **7.** 180 gallons

Page 75   **1.** $150.00, $138.85   **2.** $190.00,
$176.25   **3.** $52.75, $46.00   **4.** $117.80, $110.88
5. $203.50, $187.95   **6.** $109.50, $101.22   **7.** 210,
$22.86, $823.55, $62.40, $761.15   **8.** 35   **9.** $3.81
10. $137.26   **11.** $10.40   **12.** $126.86
13. $761.15   **14.** $126.86

Page 76   **1.** 1164.8 miles   **2.** 2096.64 miles
3. 5824 miles   **4.** 4775.68 miles   **5.** 110 gallons
6. 140 gallons   **7.** 150 gallons

Page 77   **1.** $194.25, $181.05   **2.** $167.20,
$157.05   **3.** $180.50, $169.75   **4.** $178.00,
$166.85   **5.** $142.35, $134.43   **6.** $174.96,
$164.05   **7.** 228, $27.36, $1037.26, $64.08,
$973.18   **8.** 38   **9.** $4.56   **10.** $172.88
11. $10.68   **12.** $162.20   **13.** $973.18
14. $162.20

Page 78   **1.** 793.8 miles   **2.** 2646 miles
3. 1243.62 miles   **4.** 2196.18 miles
5. 120 gallons   **6.** 130 gallons   **7.** 160 gallons

Page 79   **1.** $114.70, $107.95   **2.** $126.36,
$119.41   **3.** $76.80, $70.15   **4.** $193.05, $181.80
5. $190.00, $176.88   **6.** $135.36, $124.94
7. 223, $22.20, $836.27, $55.14, $781.13   **8.** 37
9. $3.70   **10.** $139.38   **11.** $9.19   **12.** $130.19
13. $781.13   **14.** $130.19

Page 80   **1.** 1640 miles   **2.** 2591.2 miles
3. 2722.4 miles   **4.** 3739.2 miles   **5.** 110 gallons
6. 130 gallons   **7.** 160 gallons